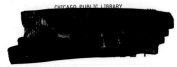

The THINK Series

THINK about
Space

The titles and authors in the THINK Series include:

SPACE: Where Have We Been and Where Are We Going?
by Isaac Asimov.
IMMIGRATION: Diversity in the U.S.
by Leon F. Bouvier, Vice President of the Population Reference Bureau.
FOREIGN POLICY: The U.S. and the World
by James A. Cooney, Director, McCloy Scholars Program, John F. Kennedy School, Harvard University.
DRUGS AND SOCIETY: Responding to an Epidemic
by Richard A. Hawley, author of The Purposes of Pleasure *and* A School Answers Badk.
TEENAGE PREGNANCY
by Cathryn Jakobson, festure writer for New York, Seventeen, *and other national magazines.*
TERRORISM: The New Warfare
by Terrell Arnold, Executive Director, Institute on Terrorism; and Moorhead Kennedy, Director, Council for International Understanding, and a former Iranian hostage.
OUR POPULATION: The Changing Face of America
by Charles B. Nam, Professor, Center for the Study of Population, Florida State University; author of Population and Society.
NUCLEAR ARMS CONTROL: Understand ing the Arms Race
by Richard Smoke, Bancroft Prize Winner and Director, Center for Foreign Policy Development, Brown University.
POVERTY IN THE U.S.: Policies and Problems
by Daniel Woods, author and winner of the 1986 award for Excellence in Teaching at Boston College; and John Williamson, Chairman of the Sociology Department, Boston College.
OUR RIGHTS: Civil Liberties in the U.S.
by Reginald Wilson, Director of Minority Affairs, American Council on Education.

THINK *about*
Space

Where Have We Been and Where Are We Going?

*Isaac Asimov and
Frank White*
The THINK Series

Walker and Company
720 Fifth Avenue
New York, NY 10019

First published in the United States of America
in 1989 by the Walker Publishing Company, Inc.

Published simultaneously in Canada by Thomas Allen & Son
Canada, Limited, Markham, Ontario

Library of Congress Cataloging-in-Publication Data

Asimov, Isaac, 1920–
    Think about space : where have we been and where are we going? / Isaac
Asimov and Frank White.
        p.   cm.—(The Think series)
    Bibliography: p.
    Includes index.
    Summary: An overview of the history and present and future role of human
beings in space tracing the development of space exploration and discussing the
challenges of the present day and the possibilities of the future.
    ISBN 0-8027-6766-4.—ISBN 0-8027-6767-2 (pbk.)
    1. Space sciences—Juvenile literature. [1. Space sciences.]
I. White, Frank, 1944–   . II. Title. III. Series.
QB500.22.A84   1989
500.5—dc19   88-36731
                                                                                    CIP
                                                                                    AC

The THINK Series Editors: William N. Thorndike, Jr.,
Ramsey R. Walker

Reading consultants: Anne-Marie Longo, Paula Sable
Chief researcher: Rachel Strauss
Jacket design: Joyce C. Weston
Text design: Joyce C. Weston
Photo research: Diane Hamilton
Jacket illustration: Tom Hughes
Text illustrations: Jeff Danziger
Appendix A written by J. B. Schramm

The editors would like to thank the many teachers, librarians, and students who
assisted in putting together the THINK Series. It would be impossible to thank
everybody; however, we would especially like to thank the following people: John
Buckey, Betty Carter, Jim Davis, Mike Hartoonian, Tedd Levy, David Mallery,
Mike Printz, Bill Polk, Ellen Ramsey, Mary Tabor.

Figures 3, 4, 5, 6, 9, 12 courtesy of the Library of Congress; Figures 10, 13, 14, 15,
16, 17, 18, 19, 20, 21, 23, 24, 25, 26, 27, 28, 33 courtesy of The National Aeronautics
and Space Administration; Figure 8 courtesy of the National Air and Space
Museum, Smithsonian Institution.

Acknowledgments

Acknowledgments

I would like to acknowledge all the members of my immediate family for their support of my work. My son, Joshua White, deserves special mention because he read and critiqued a major portion of this manuscript for me.

Frank White

Contents

THINK about
Space

1 | Introduction

- A bearded man dressed in animal furs blows on his hands to stay warm as he watches a caribou in the distance. The animal crosses a narrow strip of land spanning an ice-covered bay. The man has never seen the country on the far side. Should he go after the caribou, into the unknown? The tribe is hungry; he picks up his spear and follows.

- A man leans down from the crow's nest of a sailing ship tossing in the waters of the Atlantic Ocean. He points to the western horizon and shouts, "Land ho!" The captain of the ship, one Christopher Columbus, smiles with grim satisfaction. "I was right," he says to himself.

- A man dressed in a bulky white suit clambers down the ladder of a spiderlike space vehicle. As his foot touches the dusty soil beneath, he says, "That's one small step for a man; one giant leap for mankind." Millions of human beings watch and listen from 240,000 miles away, on Earth. Neil Armstrong begins his "moonwalk."

The lives of these men were separated by thousands of years. One of them was what we would call a "caveman." One of them was a sailor on the ship that brought Christopher Columbus to the "New World." One of them was an astronaut, the first man on the moon.

These people were all different, but they were united by one thing—they were explorers. One was trying to find new food for his tribe, another to bring home the wealth of India, and another to be the first on the moon.

Still, they all showed a human desire to explore the environment. Human beings have probably always been explorers. Scientists think that we filled up the planet Earth by starting out in one place (many think it was East Africa) and then wandering from conti-

*Figure 1—A sailor scans the horizon for land from the crow's
nest of Christopher Columbus's ship, the **Santa Maria**. Since the
Stone Age, man has felt an irresistible urge to explore the world
around him. This urge is manifested today in our exploration
of space.*

nent to continent until there were human beings
everywhere—even Antarctica.

THE FINAL FRONTIER

During that time, there was always a new frontier on
Earth. For Americans, it might have been the West;
for Russians, it was Siberia; and for Brazilians, the
Amazon. Some think that the oceans and polar
regions are still frontiers.

However, planet Earth has been pretty thoroughly explored now, and for many, there is only one frontier worth considering—space, "the final frontier" of "Star Trek" fame.

When we stop to think about it, the current time is a truly unique period in history. It is the time when human beings are beginning to establish a permanent presence in outer space. It is like the moment when the Pilgrims first set foot on Plymouth Rock in 1620 and began to explore and develop the NEW WORLD of North America.

As at the beginning of America, it is a good time to reflect on what human beings have done and will do in outer space. Will we make the same mistakes we have made on Earth? Will we fight wars on Mars and degrade the environment of the moon? Or will we learn new ways of thinking and acting that will help us to begin anew, creating different kinds of human communities in outer space?

LIVING IN THE SPACE AGE

Since 1957, human beings have been living in a new age, the Space Age. That was the year when the Soviet Union launched the first artificial Earth-orbiting satellite, called *SPUTNIK*.

While just over 200 people have actually left the planet Earth and gone into space, all of our lives have been affected by space exploration and development. Without the hundreds of satellites that now ring our planet, today's sophisticated weather forecasting would be impossible. Also, much of the news and entertainment that we take for granted would not be available without communications satellites. In the words of Marshall McLuhan, an expert on the media who wrote extensively about television in the 1960s and 1970s, satellites have truly made the Earth a GLOBAL VILLAGE.

Most of us now realize too that computers are likely to become an important part of our educational or work lives. What we may not know is that today's small, fast computers are children of the American space program. In 1961, when President Kennedy committed the United States to put a man on the moon and bring him back safely within the decade, we didn't have computers that could do the job.

It was the effort to fulfill the president's order that resulted in the development of smaller, faster computers for the Apollo spacecraft. The heritage of today's Apples and IBM personal computers begins with the Apollo program.

Beyond the many material benefits of space exploration, of course, there are other valuable "spinoffs." Scientists are learning more about the universe every day, from manned space missions, unmanned robot "probes" to the planets, and even from telescopes in orbit around the Earth.

We are learning more about ourselves too. Pictures of the Earth from space show a beautiful blue marble floating in the blackness of space, giving us a new sense of how we fit into the universe.

Now human beings are poised to begin the next phase of our adventure in space. Are we ready for it?

THE ISSUES

We have shown that we can explore outer space, even live there. Now, for many people, the question is *should* we do it, and if so, how? Some feel that the money spent on exploring and developing outer space would be better spent solving problems on Earth.

Others believe that space exploration will become just another competition between nations, driven by desire for military advantage. Equally sincere and thoughtful advocates of space exploration believe that we must leave the planet to fulfill our destiny. They also argue that we can only solve Earth's problems by getting outside the Earth and gaining a new perspective.

What do you think? Young people in junior high and high school today will be adults in the year 2000. They will be responsible for the decisions that will shape the future of humans in space. Some will be rocket scientists, engineers, astronauts, or even space settlers. But all human beings, no matter what their roles, will be affected, and should be informed about the issues.

THE FUTURE

Where might the exploration of space lead us? The question is simple, but the answers are complex. We are just not used to thinking about the possibilities that the entire universe offers us. Our planet, which seemed to be limitless to our early ancestors, is truly tiny in the universal scheme of things.

Just imagine this: scientists say that there are about 200 billion stars in our own MILKY WAY GALAXY alone! They also estimate that there are about 100 billion galaxies in the universe! The nearest star is over four light years away. This means that light, which travels at 186,000 miles per second, takes over four years to get from there to here.

Our own galaxy is tens of thousands of light years across, and we know about galaxies that are hundreds of millions of light years away. What if all the human beings alive today devoted the rest of their lives to exploring space? We would only barely get started on the job.

Exploring outer space opens up a limitless future for us. There is no way to imagine where it might lead. However, choices that we make today will affect that future. Also, these choices do not just affect our future in space, but on Earth as well. Our decisions will have an impact not only on our lives but on those of our children, and their children too.

IT'S UP TO YOU

Where do you stand on the question of space exploration? Do you think we should be doing more or less? Do you think about it a little bit or a lot?

This book will challenge you to think about space exploration, where we've been, and where we're going. Chapters One and Two talk about where we've been, in the early years and during the dawn of the Space Age. Chapters Three and Four consider where we're going, looking at the issues that human beings of today and tomorrow must confront. In the Appendices, you'll find more information on how people used to think about space, as well as career opportunities for the future. As you read along, just remember that if you're well informed, you can do more than express your opinion. You can make things happen on the final frontier.

REVIEW QUESTIONS

1. What is similar about a caveman, a sailor, and an astronaut?
2. Where is today's frontier? Why?
3. Why did the "Space Age" begin in 1957?
4. What are some of the issues that confront us in exploring outer space?
5. Why is it important for students to be informed about space exploration?

2 | The Early Years

SPUTNIK SPEAKS

On October 4, 1957, the people of planet Earth heard a new sound coming from outer space. Beep . . . beep . . . beep . . . beep. No, it wasn't a signal to Earthlings from aliens on another planet. The first sounds to be heard from space came from a vehicle produced on Earth. It was an artificial satellite called *Sputnik* (a Russian term meaning "fellow traveler of Earth").

A new sight also appeared in the night sky. With a keen eye, the Earth's new "moon" could be seen passing overhead. For the first time in history, human beings had put something from Earth into space. The Space Age had begun.

The people of Earth immediately recognized the Soviet Union's incredible achievement in launching *Sputnik*. Some, especially in the United States, reacted with shock. They had thought the Russians were very backward technologically, and that the United States would be first in space.

Like many events in history, the launching of *Sputnik* seemed to happen all of a sudden. People were not prepared for it, and that's why it was such a surprise. Perhaps they also knew that they could never again think about space as they had in the past. Everything was new and different now.

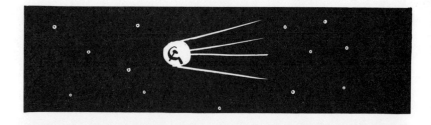

Figure 2—The launch of the Soviet satellite Sputnik *in 1957 worried a great many Americans. Would the Soviets dominate space? Would this tip the fragile Cold War balance of power? The country responded by devoting substantial resources to catching up with the Soviets.*

However, *Sputnik* should not have been a shocking event for anyone. Many small steps had been taken over the years leading up to the launch, and the Russians had been quite open about their intentions.

To understand the surprise that greeted *Sputnik*, it is necessary to go back even before 1957 and see how people have thought about space, and Earth, through history.

EXPLORING SPACE FROM EARTH

Sputnik was not the beginning of human exploration of outer space. It just seemed that way at the time. The truth is that human beings have been exploring outer space for thousands of years. You are exploring outer space when you look up at the night sky and locate a constellation or a planet. Astronomers do the same thing, with more sophisticated equipment, and they have been doing it since civilization began.

In ancient times, it was important for people to understand the heavens for practical reasons. They needed to know the seasons, so that they could plant their crops at the right times. For those who believed that the stars and planets affected human lives, a knowledge of the universe was also critical. For thousands of years, people did not separate astrology (the supposed effect of stars and planets on individual lives) from astronomy (the study of stars and planets unrelated to their effects on human beings).

As they gathered information from the sky, our ancestors built pictures of how the universe was constructed. These views of the universe form a branch of knowledge called COSMOLOGY (knowledge of the cosmos or universe). Until the 1500s, most people imagined the Earth to be the unmoving center of the universe. They believed that the stars and planets revolved around us, while we stayed put.

To some religious leaders, the heavens and the Earth were fundamentally different. What was in the heavens was perfect, and what was on the Earth was "imperfect" or "fallen."

Of course, no one could leave our planet at that time, which didn't help the various cosmologies to become more accurate. In addition, before the telescope was invented, all observations were made with the naked eye. That limited what could be seen.

However, there were several sticky problems with the picture of the universe that put the Earth at the center. One of the biggest problems was that it made it incredibly difficult to calculate the orbits of the other planets.

As seen from Earth, the planets did some crazy things. They seemed to stop in their orbits, back up, turn around, and then go back in the same direction again. We know today that they only appear to do this. Why? Because the Earth is also moving, and the planets appear to go backward as we catch up with them and pass them on our way around the sun. (If you think about it, the same thing happens when you pass another car on the road.)

To account for this behavior, astronomers of the Middle Ages had to draw strange patterns called EPICYCLES that showed the other planets leaving their orbits and doing "loop-the-loops" now and again.

A European astronomer named Copernicus (1473–1543) was frustrated by the fact that even epicycles did not explain all his problems with planetary orbits.

What to do? He knew that some people in the past had suggested that the Earth and the other planets actually went around the sun. He started to think about how the solar system would look if he thought about space in that way.

When Copernicus assumed that all the planets went around the sun, the calculations were still not quite right, but a giant step had been taken. Copernicus's discovery has been called the COPERNICAN REVOLUTION. It was not a revolution with guns that

Figure 3—Nicolaus Copernicus, shown here with T-square, compass, and other astronomical tools, hypothesized that the sun, not the Earth, lay at the center of the solar system.

overthrew a government. It was a revolution in how people thought about space, the Earth, and themselves.

Copernicus was followed by Kepler (1571–1630), who worked out the exact mathematics for describing

Figure 4—An illustration of the Ptolemaic concept of the universe, with the earth at the center. This theory was popular with religious groups because it confirmed humanity's importance in God's plan (note Christ and the Apostles at the top of the illustration).

the planets' orbits around the sun. Kepler also corrected Copernicus's major mistake. He found that the planets did not go around the sun in a circle, as Copernicus had thought. They went around the sun in an ellipse or egg-shaped pattern. Now, the calculations really worked.

Kepler's mathematical descriptions of the planets' orbits filled out the new perspective on the solar system that had been suggested by Copernicus. Slowly but surely, a new way of thinking about the universe was emerging.

Copernicus and Kepler did their work without the benefit of the telescope. They were still limited to observations that they could make with their own eyes. Then, another breakthrough occurred when Galileo (1564–1642) developed one of the first telescopes and turned it to the sky. All of a sudden, new objects appeared in the sky that had never been seen before! Imagine how Galileo felt when he aimed his telescope toward the giant planet Jupiter and saw four moons revolving around it! No one had ever seen these moons, or even knew they were there. Galileo also studied the Earth's moon and other parts of the solar system in some detail.

He made many other discoveries, and the new information convinced him that the view of the universe advanced by Copernicus and Kepler was correct.

How important was it that people started to change how they thought about the solar system? This new perspective was highly controversial, especially with the Catholic church. The leaders of the church forced Galileo to take back his statements supporting the other astronomers. They held to the old view of the

Figure 5—Galileo Galilei was the first man to see the moons of Jupiter. His observations confirmed the controversial theories of Copernicus and Kepler.

Earth at the center, and they felt threatened by any ideas that contradicted that image. They felt that humanity's importance in God's plan would be diminished if the Earth were found to be just another planet revolving around an ordinary star. However, the change in thinking had begun, and the church couldn't stop it, no matter how hard it tried.

PERSONAL PROFILE

Sir Isaac Newton and the Laws of Nature

Isaac Newton was born in 1642, the year Galileo died. He liked to do experiments as a boy, which sometimes distracted him from his schoolwork. After a slow start at school, he soon showed his brilliance and was sent to Cambridge University. In the first few years after his graduation, he worked out the universal law of gravitation.

We are told that it was Newton's curiosity about falling apples that led him to discover the laws of gravity. Watching an apple fall from a tree, he asked himself the question, "Why doesn't it fall up instead of down?" Whether the story is exactly right or not, his creative ideas changed our views of the universe forever.

Newton made many contributions to the development of science. However, his discoveries of universal "laws of nature" are among the most important. What if it had been true, as many believed, that the "heavens and the Earth" were really different places? That would mean that you couldn't count on a spaceship behaving the same way when it got to the moon as it did when it first left the Earth. Instead of falling toward the moon when it approached it, the ship might be repelled away from the lunar surface!

If the laws of nature changed from planet to planet, the way human laws change from country to country, it would be difficult to plan or implement a space program. We couldn't say, "Let's send a spaceship to the moon, or to Mars," because we would have no idea what would happen when it got there. Who would agree to put money into such an uncertain effort?

However, Newton was able to show that there are indeed universal "laws." He showed first that the LAW OF GRAVITY can be applied anywhere. He also formulated the three LAWS OF MOTION, including the third: "For every action, there is an equal and opposite reaction." When a rocket fires, it sends gases out of its tail. That is the action. When the rocket moves in the opposite direction from this thrust, that is the reaction.

Taken together, these two laws are basic to any program of space exploration. The third law tells us how our spaceships can move about, and the law of gravity tells us how that motion will be modified by the pull of planetary bodies. These and other contributions by Sir Isaac Newton were enormous. In fact, the "Newtonian view of the universe" was the dominant one until the twentieth century, when Einstein's work showed us Newton's limits.

During one period of his life, Newton behaved quite strangely, and some people believe that he had a mental illness. However, he seemed to recover toward the end of the seventeenth century. He was elected to Parliament and became the head of the British Mint. He held the post at the Mint for twenty-eight years.

Newton's statement about his work has been quoted often because it clearly summarizes the search for knowledge. If he had done anything significant, he said, it was because "I have stood on the shoulders of giants."

The work of Copernicus, Kepler, and Galileo was aided by the efforts of Sir Isaac Newton (1642–1727). His famous encounter with a falling apple led him to determine how the laws of gravity worked. These laws appeared remarkable to people at the time, because they seemed to hold true on Earth and in the heavens. "Heaven and Earth" seemed to be more alike than ever before.

These are just a few of the people who explored the universe without leaving Earth. Before them, there were also great astronomers among the Maya people of Central America, the Chinese, and in the Middle East. Today, there are still astronomers and other people who spend their lives thinking about space without leaving the planet.

When the Apollo astronauts went to the moon, they relied on the work of all those earlier thinkers. For example, without the knowledge of Newton's laws of gravity and motion, they would not have known how to navigate to the moon.

It is natural, of course, that where people can go in their minds, they would like to go physically. For that reason, building spaceships and leaving Earth has long been a human desire. It took just under 400 years from the time Galileo's eyes left the Earth to look at the moon to the time that Apollo astronaut

Neil Armstrong became the first person to walk on the lunar surface.

During that 400-year time period, many things happened to take the human body where the human mind had already gone. The advent of rockets was probably the most important development in extending the field of space exploration from the surface of Earth out into space itself.

ROCKETS AND SPACESHIPS

Rockets are important in discussing space exploration because without rockets it would be difficult if not impossible to leave the planet Earth. All of our "space exploration" would be like that of the early astronomers, limited to what we can see from here.

Perhaps someone will eventually invent an "antigravity" device, or a "transporter" such as they have on "Star Trek." Until then, the only way we know to get off our planet is to fill up a big cylinder (spacecraft) with fuel and oxygen and then "light a match" down at the bottom.

The resulting burning pushes down against the Earth with tremendous force. That in turn pushes the spacecraft up and away from the Earth. If there is enough power behind the push, *bingo,* you find yourself in outer space in about eight minutes.

Once again, we see the workings of one of Newton's universal laws. Because these laws of nature are always in force everywhere, the explosion of a rocket can be used to get astronauts or satellites into space and back again.

In addition to being good astronomers, the Chinese are credited with inventing rockets during the elev-

Figure 7—The Chinese invented rockets in the twelfth century for use in warfare. These primitive rockets were often as dangerous for the Chinese as they were for their enemies.

enth century. These early rockets were powered by gunpowder, which the Chinese are also said to have invented.

As with many new inventions, rockets saw their first use in warfare. During battles, the Chinese would light the rockets and fire them at the enemy's horses. The horses, understandably enough, found rockets to be a frightening addition to the battlefield and usually ran away.

However, those early rockets could not be controlled, and they often caused as many problems for the Chinese troops (and horses) as for their enemies. Many good inventions are used soon after they are developed and then abandoned because of technical problems that no one can solve. This is what happened with rockets.

Rockets are still with us in a simple form called "fireworks." Today's "bottle rockets" aren't much different from those used by the Chinese in their wars. Fireworks weren't enough for some people, however. They kept working on rockets, both for military and peaceful purposes. Some of them also saw that if a rocket were big enough and powerful enough, it could be used to lift objects and people into outer space.

Having a way of getting into outer space isn't enough, however. The question was, and still is, not just whether we can leave the planet, but also whether we should. People want to know, "Why should we go into outer space?" The history of thoughts about this question is as important as the development of rockets.

Russians have been especially interested in answering these questions and have been thinking about space exploration for a long time. For example, in the mid-nineteenth century, Nikolai Fyodorov inspired listeners with his discussions of humanity's "cosmic destiny." He believed that humanity's natural habitat was space, not Earth. He thought that humans would someday play a major role as regulators of natural processes throughout the universe. Fyodorov was not a professor but a librarian who worked for most of his life at the main library in Moscow. But his ideas are important even today to the Soviet space program.

A Russian teacher named Konstantin Tsiolkovsky built on Fyodorov's ideas and proposed a multistep "space program." Tsiolkovsky's plan included many of the elements that are common to today's space programs, and he believed that space exploration would eventually lead to the perfection of humanity.

Tsiolkovsky, like Fyodorov, was an ordinary person with an ordinary job, but his thoughts about space exploration still have great impact on people. In fact, there is now a statue of Tsiolkovsky in front of the Cosmos Hotel in Moscow. Following in the footsteps of people like Copernicus and Kepler, Tsiolkovsky and Fyodorov explored space without leaving Earth. Yet they went so far that the American ASTRONAUTS and Soviet COSMONAUTS are still catching up with them!

The work of Fyodorov and Tsiolkovsky inspired many scientists in Russia to become interested in space exploration. When the Russian Revolution took place in 1917, the new Communist government supported this interest. Communist thinking supported technological development because it was important in making the Soviet Union more modern. For this reason, the government began to support developments in fields such as rocketry.

Individuals in other European countries, especially Germany, started "rocket clubs." These clubs maintained an interest in rockets and space exploration at a time when there wasn't a lot of government money available to support the work.

In the United States, an inventor named Robert Goddard launched the first liquid-fueled rocket in

Figure 8—Rocket hobbyists tinker with an early liquid-fueled rocket near Stockton, New Jersey, in 1932. Private individuals and clubs helped to sustain interest in rocketry before government funds were available for research.

1926. Goddard is another good example of how dedicated individuals made a major difference in the field of space exploration. He worked almost alone in a small town in western Massachusetts, and few expected him to succeed in realizing his dreams. But Goddard's contribution to space exploration was very significant, and the reason goes all the way back to those Chinese rockets. From the days of the Chinese until Goddard's time, all rockets were "solid fueled." This meant that they used solid substances such as gunpowder to create their power.

The problem with solid-fueled rockets, as the Chinese found, is that their burning cannot be controlled once they are lit. This problem was described many times after the explosion of the space shuttle Challenger in 1986. That accident was caused by a failure in one of the solid rocket boosters that lifted the shuttle off the pad. Once the boosters started burning, the people in Mission Control could not stop them except by exploding them with a remote control device. By then, of course, it was too late.

With a liquid-fueled rocket, the burning can be controlled. The fuel is kept in storage chambers and then pumped into firing chambers, much like an automobile engine. This control gives rocket engineers a lot more flexibility in designing and operating rockets and spaceships. In fact, when problems have developed with the liquid-fueled engines on the space shuttle, engineers have been able to shut them down minutes before a launch.

There have been many changes in rocket design since the 1920s. However, the *SATURN V* that took men to the moon works on the same basic idea as Goddard's first rocket. Today, even as engineers work

Figure 9—Dr. Robert H. Goddard, pictured here in 1926 with one of his early rockets, was the first engineer to propel a rocket with the steadier burning liquid fuel.

on new designs for spaceship engines, the rocket is still the workhorse taking human beings into space.

SPACE AND THE HUMAN SPIRIT

Technology—rockets, spaceships, computers, and so on—is the way we get into space. However, the drive to go there has always come from people and their dreams. There will always be a modern equivalent of the caveman who is willing to cross the ice floes in search of new game, or a sailor who will sail the uncharted ocean looking for new wealth. In the words of "Star Trek," there will always be people who "boldly go where no one has gone before."

Most of the people who built rockets didn't do it just to see if rockets would work. Often, they were lured on by a dream of space exploration. For example, in the late nineteenth and early twentieth centuries, the idea of going to Mars inspired both Tsiolkovsky in Russia and Goddard in the United States. The story is told that the young Goddard was sitting in a cherry tree one evening in 1899 when he started thinking of Mars. Then he imagined a spaceship that could take him there. These thoughts inspired him to do the work that resulted in the liquid-fueled rocket.

Interest in space exploration and rocketry was especially strong during the 1930s in Germany. This was the birthplace, for example, of Wernher Von Braun, who eventually directed America's APOLLO moon program. Like Goddard and Tsiolkovsky, he was drawn to space exploration by more than an interest in rockets. Von Braun could imagine human beings exploring, living, and working in space. Von Braun was also a practical person. He had an ability

Figure 10—Dr. Wernher Von Braun, NASA's rocketry wizard, engineered the rockets that eventually allowed the United States to reach the moon in July 1969.

to organize people and he learned how to build large rockets.

One person, like Goddard, can build a small rocket that is launched from an icy pond in Massachusetts. One person, like Tsiolkovsky, can write up his ideas about exploring space. However, one person could hardly build a moon base or a spaceship that would go to Mars. In the 1930s, governments began to provide support to the early space pioneers. This change led people into two very different ways of thinking about space: as a new environment in which peace could be established among nations, or as a new "high ground" from which to wage war.

BIRTH OF THE MISSILE

As in the time of the Chinese (and today) rockets interested governments because they could be used as weapons. During World War II, many scientists were organized into teams working on the use of rockets in the war effort. Von Braun and a group of German scientists working at a base called Peene- münde were especially successful. Toward the end of the war, their V-2 ROCKETS were raining destruction on London and other cities in England. The rocket had become a missile.

If the German armies had not already been close to defeat, the V-2 might have made a big difference in the war. But it was too late. In 1945, the American and Russian armies invaded Germany from two dif- ferent directions. Both sides knew that missiles would be an important weapon of the future. Both sides wanted the German rocket scientists too.

Von Braun decided that if he were going to be

captured, he wanted to be captured by the Americans. He and many of his colleagues began to filter toward the approaching American armies. They succeeded, and were eventually brought to the United States, settling at the Army's Redstone Arsenal in Huntsville, Alabama.

The Russians captured other German scientists and took them back to the Soviet Union. As "prizes of war," the two groups of scientists helped the Americans and the Russians dominate world events for the next forty years. As missile builders, they helped fight the COLD WAR. As spaceship builders, they helped create the "Space Age."

The birth of the missile and of the atomic bomb came at almost the same time. In fact, one of the reasons that American scientists worked so hard on the atomic bomb was the fear that the Germans would invent it first. The idea that Adolf Hitler could put the bomb on the V-2 and deliver it to London was too terrifying to consider.

Hitler failed to do that, but now it can be done. As many people have noted, the rocket can now be used to destroy everyone on Earth, or to free us all from Earth.

The rocket doesn't decide what it does. The people who make rockets do. That's one reason that thinking about space has become so important to us today. It has a big effect on whether we use rockets as missiles or spaceships.

REVIEW QUESTIONS

1. In what ways did human beings explore space before the rocket became available as a spaceship?

2. How has thinking about space changed our views of the universe?
3. How are Galileo and Neil Armstrong similar?
4. Would you rather be an astronomer or an astronaut? Why?
5. Why wasn't the rocket used as a spaceship when it was first invented?
6. Why did the Russians build a statue of Tsiolkovsky in Moscow?
7. How did Robert Goddard change the direction of space exploration?
8. How was Wernher Von Braun different from Goddard?
9. What is the difference between a missile and a spaceship?

3 The Dawn of the Space Age

LAYING THE FOUNDATIONS (1945–1957)

World War II ended in 1945, just twelve years before the launching of *Sputnik I*. The people of planet Earth didn't know it, but the world would soon change dramatically and forever, as the foundations of the Space Age were laid in those dozen years. During that time, astronomers continued to study the universe from Earth's surface. The United States and Soviet Union continued to develop the missile as a weapon of war.

However, something else was happening at the same time. The idea of the rocket as a spacecraft began to move out of the realm of science fiction and into the minds of scientists. The Space Age did not really dawn until the missile became a spacecraft. This allowed human beings to put their instruments and themselves into space. That was the real significance of *Sputnik*.

From 1945 until 1957, there was not much change in how people thought about space exploration.

In 1946, the writer and scientist Arthur C. Clarke first described the potential benefits of COMMUNI-CATIONS SATELLITES in orbit around the Earth. However, the technology to realize his concept did not yet exist, and most people probably saw his paper as a "think piece." How could they imagine that a satellite called *Sputnik* would be beeping down at them in nine years?

In 1948, the astronomer Fred Hoyle said that if we could ever get a look at the Earth from space, it would start a revolution in our thinking. Hoyle understood that we would then see our home planet as a whole, and gain a new understanding of our place in the universe. He was right, as the astronauts and cosmonauts later discovered, but that seemed a long way off at the time. People were probably more concerned with getting a look at the new 1948 automobiles than in seeing the Earth from space.

In the United States and the Soviet Union, rocket research went on. The rocket also continued to have two faces, one friendly, the other hostile. While some scientists worked on ways to use rockets in creating new knowledge, others looked at how they could be used to destroy cities, towns, and military installations.

During the 1950s, the United States and the Soviet Union competed for world power. They alone had nuclear weapons and the airplanes and missiles to deliver them anywhere in the world. However, neither the United States nor the Soviet Union dared to use these weapons in a war. Both of them knew that a nuclear war would result in the utter destruction of both sides.

For this reason, the United States and Soviet Union did not have any "hot wars" from 1945 to 1957. American and Soviet soldiers fired no shots at one another during that time. Instead, they began a Cold War in which they competed with each other for the allegiance of other nations.

The politics of Earth were much more important than space exploration to the nations of Earth in the 1950s. Of course, even at the height of the Cold War, the United States and the Soviet Union competed in some areas and cooperated in others, just as they do today. Cooperation was supposed to be the idea behind the INTERNATIONAL GEOPHYSICAL YEAR (I.G.Y.) of 1957, a multinational project designed to study the Earth. The United States and the Soviet Union both announced their intentions to launch an Earth-orbiting satellite during that year.

Thus, the stage was set for the "shocking" moment when *Sputnik* began to speak to the people of Earth from outer space.

THE SPACE AGE BEGINS (1957)

Everything seemed ordinary enough on that last day before the beeping of *Sputnik* was first heard. On October 4, 1957, the *New York Times* carried a front-page story on the World Series game between the New York Yankees and the Milwaukee Braves. It included a big picture of Wes Covington of the Braves making a "spectacular backhand catch" of a line drive hit by Bobby Shantz. The Braves went on to win the game and the series.

The World Series game was the big story that day, but there were hints of things to come. Tucked just to the left of the picture of Wes Covington making his catch was a story by reporter Walter Sullivan under the headline, "Soviet Expert Tells West of Test Rocket."

The story said that Russian scientists gave details of a Soviet rocket to an "astonished audience" at a meeting of the National Academy of Sciences in Washington. The story also said that the rocket was described at a conference "to coordinate plans for rocket and satellite firing during the I.G.Y., an eighteen-month study of the Earth that began July 1."

The rocket described at this meeting was apparently not the one used the next day to put *Sputnik* in orbit. However, the scientists did discuss satellites. For example, the Soviet delegation asked that the American satellites, when they were launched, use their sophisticated cameras to take pictures of the Russian satellites, when they were launched.

On this last day before the Space Age began, Russian and American scientists were talking openly in Washington, D.C., about launching satellites within the next several months. There was no feeling of urgency or competition, however.

On the next day, though, things changed. A giant rocket lifted off the launch pad of a base somewhere in central Russia. It was a descendant of the horse-frightening Chinese rockets of the Middle Ages. It was also a child of Robert Goddard's little liquid-fueled rocket of more recent times.

That rocket also carried with it the dreams of people such as Tsiolkovsky, Goddard, and Von Braun. Perhaps the dreams of Kepler, Copernicus, Galileo, and Newton were realized as well.

On October 5, 1957, the day following *Sputnik*'s historic launch, the reports about Russian space efforts were not so calm. Now the headlines ran all the way across the page, screaming that the Soviet Union had put a satellite in orbit. The World Series faded into the background, and the papers were full of news about the Soviets' achievement.

The reaction of the United States can only be described as complete shock. The basic questions were "Why did the Soviets beat us into space?" and "Why didn't we launch the first satellite?" The success in space shed a new light on the politics of Earth. The United States and the Soviet Union had been trying to convince the other nations that each had the "best" society. If the Soviet Union were first in space, didn't that mean they were the best?

The debate may not make much sense to us thirty years later, but it was taken very seriously at the time. Many influential people felt that the United States had fallen behind the Soviet Union, and that much had to be done to catch up.

When *Sputnik I* was launched, President Eisenhower tried to respond calmly. His administration said that *Sputnik* did not change the military situation, and that there was no need to overreact to the Soviet success.

A month later, on the fortieth anniversary of the Russian Revolution, *SPUTNIK II* roared into orbit, carrying the first living being into space, Laika the dog.

The Eisenhower administration now felt even stronger pressure to get an American satellite into orbit. Of course, that was easier said than done. The United States did have a satellite program called VANGUARD being developed as part of the I.G.Y. *Vanguard* was scheduled to be test-fired at CAPE CANAVERAL in Florida in December 1957. The administration decided, however, that it should become a full-scale launch attempt.

The nation watched expectantly on live television as the countdown progressed. Five, four, three, two, one . . . fire! The *Vanguard* rocket lifted a few feet off the pad, then fell backward and crumpled in a fiery explosion. The grapefruit-sized satellite bounced out of the nose cone of the rocket and actually started transmitting: beep, beep, beep, like *Sputnik* had.

Unfortunately, *Vanguard* was not in orbit, as *Sputnik* had been. It was rolling along the Earth at Cape Canaveral, while the rocket that was supposed to put it into space burned in the background.

As 1957 drew to a close, the United States viewed the dawning Space Age with deep concern. Americans now knew that the Russians had a rocket powerful enough to put big satellites in orbit. They realized that the same rocket could eventually put people into space or deliver a nuclear weapon to America.

Many prominent Americans believed that control of space meant control of the Earth. They assumed that to be the real motive behind the Russians' space efforts. Whether it was true or not, that was how they thought about space, and it had a major impact on the next ten years.

THE RACE IS ON (1958–1969)

Remember Dr. Von Braun and the German scientists stationed in Huntsville, Alabama? They had continued working throughout the 1950s on rocket experimentation. They used V-2s developed in Germany and brought to the United States after World War II, and built on the work of Dr. Goddard and other American rocket researchers.

Von Braun was pretty certain he could put a satellite into orbit if he got the chance. However, the administration wanted to emphasize the peaceful uses of outer space. There was concern that using a military rocket, descended from the V-2, would not accomplish that goal. The competing Vanguard project was actually managed by the U.S. Navy, but it used a system with far less military hardware. Ultimately, *Vanguard* was assigned the mission.

However, when the *Vanguard* failed on television, in full view of the world, the government decided it was time to try something else. The job was turned over to Von Braun and his team toward the end of 1957. On January 31, 1958, they successfully launched the *EXPLORER* satellite, using a modified Jupiter-C rocket.

On March 17, 1958, the *Vanguard* was finally launched successfully as well. Americans felt that they were beginning to achieve equality with the Russians in space. Now, they said, the race is really on! When was the term "space race" first used? It is probably impossible to answer that question accurately. However, the people of the United States certainly started to see space exploration as a race after *Sputnik* was launched and *Vanguard* failed.

Was it good or bad to think of space exploration as a race? That issue is still being debated, but it certainly had a dramatic effect on the United States. For example, schools all over the country changed their curricula almost overnight. Courses were made harder, and administrators put a new emphasis on teaching science and mathematics.

The "space race" idea also accelerated the American space program. Without the Russian challenge, it seems likely that the United States would have approached the space frontier much more gradually.

People assumed that the Russians were "ahead in

Figure 11—During the 1960s, there was intense competition in space between the Soviet Union and the United States. The media dubbed this competition the "Space Race."

the race" because the Soviet school systems produced more scientists and engineers. No one had heard of Fyodorov and Tsiolkovsky (who weren't scientists or engineers), so they didn't realize that the Russians had been thinking about space exploration for a long time. Very few Americans had ever heard of Goddard, their fellow countryman. They didn't know that he had already pioneered many of the techniques that were needed to build advanced rockets and spaceships.

In addition to launching its first satellite in 1958, the United States also put all space activities under one government agency—the NATIONAL AERO-NAUTICS AND SPACE ADMINISTRATION (NASA). Legislation passed in that year gave NASA overall responsibility for the nation's exploration of space, emphasizing the peaceful uses of the new environment.

With NASA in place, the United States also moved toward manned spaceflight—but not without debate. Many scientists argued that humans weren't needed on spaceships. Unmanned rockets could do the same thing for less money, they said. Others felt there would be times when having a person on board would help a mission to succeed. They also believed that public support for space exploration would diminish without manned flights.

The Eisenhower administration decided to go ahead with a mixed program of manned and unmanned flights. Then another intense debate took place inside the government, because no one knew what kind of person would make the best astronaut. After all, there had never been an astronaut in human history!

Finally, President Eisenhower brushed aside suggestions that gymnasts, circus acrobats, or other risk-takers might be just right for the job. Military test pilots, he said, would be America's first men in space.

In 1959, NASA selected the first seven test pilots as astronauts for the Mercury program. The "Mercury Seven," as they came to be known, represented American hopes of recapturing the lead in space exploration.

Once again, however, the Soviet Union beat the United States to the punch. On April 12, 1961, a

Russian Air Force major named Yuri Gagarin became the first human being to leave the planet Earth. The beeps of *Sputnik* had been made by a Russian machine, and now the first language spoken in space was Russian as well.

It is almost impossible for us to imagine the feelings that Gagarin must have had when his spacecraft went into orbit. Like the caveman who courageously crossed the ice floes, or the sailor who ventured across trackless oceans, Gagarin was a true explorer. Unlike them, however, his exploration was the first to ever take a human being into orbit.

Later, Gagarin wrote about his experience in a book, *Survival in Space:*

Figure 12—Yuri Gagarin, Soviet cosmonaut, was the first man in space, in April 1961. The United States followed a month later, launching Alan B. Shepard, Jr. for a 15-minute, suborbital "hop."

Trembling with excitement I watched a world so new and unknown to me, trying to see and remember everything. Astonishingly bright cold stars could be seen through the windows. They were still far away . . . but in orbit they seemed closer than the Earth. But the point was not the distance . . . but the principle. Man had overcome the force of Earth's gravity and gone out into space.

Yuri Gagarin returned to Moscow after his single orbit of the Earth to a hero's welcome. The Soviet Union again led the space race. American prestige sank to a new low, and not only because of Soviet space triumphs. There had been political setbacks on Earth as well.

On April 17, exiles backed by the United States attempted to invade and overthrow the Cuban government of Fidel Castro, who was supported by the Soviet Union. Known as the "Bay of Pigs" invasion, the effort failed, most of the invaders were captured, and President John F. Kennedy suffered a major blow to his personal leadership.

The United States responded to these setbacks with action and words. The action: on May 5, 1961, another rocket developed at Redstone Arsenal answered the call. It put astronaut Alan B. Shepard, Jr. into space on a SUBORBITAL "hop" for fifteen minutes.

The words: on May 25, 1961, President Kennedy delivered an historic speech at a joint session of the United States Congress. He called for several different initiatives, and speaking of the American commitment to space exploration, the president made a startling declaration:

I believe that this nation should commit itself to achieving the goal, before this decade is out, of landing a man on the moon and returning him safely to Earth.

President Kennedy's bold initiative became the Apollo moon program. It helped galvanize the nation and the space exploration community. The Mercury program had barely begun and had achieved only one small success, Shepard's flight. At the time of the speech, the United States had no rockets powerful enough to go to the moon. In fact, no one was sure how to do what the president said must be done.

Figure 13—President John F. Kennedy presents Alan B. Shepard, Jr., the first American in space, with NASA's distinguished service medal in a ceremony on the White House lawn in May 1961.

However, the president's decision totally changed the situation by redefining the terms of the contest. If there is going to be a race, there must be a finish line. Now the space race had a finish line, the moon. If there is going to be a race, there must also be rules, and the president had made those up as well—to win this race, you must send a man to the moon and bring him back safely before December 31, 1969!

It is not clear whether the Russians planned to go to the moon at that time. They did send an unmanned rocket to the moon and also flew the first unmanned rocket in orbit around the moon, sending back the first pictures ever of its "dark side." Even up to the moment of the successful *Apollo 11* lunar landing, there were rumors that "the Russians are coming."

At first, the Russians seemed to remain ahead of the Americans, regardless of what the president said. After completing another orbital "hop," NASA hoped that the next flight would include many orbits, putting Gagarin's single orbital flight in the shade.

The Russians again refused to cooperate. On August 6, 1961, Russian was again spoken in orbit. "I am Eagle, I am Eagle," the voice said. Eagle was the code name for cosmonaut Gherman Titov, who orbited the Earth seventeen times in just over twenty-five hours. Titov became the first human to spend a full day in outer space.

The United States responded with the flight of John Glenn in February 1962, and Glenn became the first American to achieve orbital flight. However, the hope that his mission would establish the United States as the leader in space had been snuffed out by "Eagle."

But the situation was about to change. To imagine

Figure 14—John H. Glenn, pictured here in 1964 after his retirement from the astronaut corps. Glenn, now a United States senator from Ohio, was the first American to orbit the Earth, in February 1962.

what happened, think of a long race on the Earth, such as a marathon. In that kind of a race, we often see situations where one runner challenges another. Then he or she falls back as the challenge is answered, gathering energy for another try. Finally, with a great burst of speed, someone pushes ahead and never gives up the lead again.

In the case of the United States and the Soviet Union, the space race started to look that way from 1962 until 1968. One side would thrust with a "first" in space, and the other would respond with a first of its own, then fall back and look for another opportunity.

The Soviets put the first woman in space, and the first multiple crew. They achieved the first docking of two spacecraft in orbit, and had the first SPACE WALK. They also established an eighty-one-orbit record that looked unbeatable.

But something was happening to the American space program. The vision was beginning to take hold. The tragic assassination of President Kennedy in November 1963 deprived the space program of its visionary, but not its vision. Vice President Lyndon Johnson, who succeeded Kennedy as president, had been an ardent supporter of a strong space effort since the 1950s. He continued that support during his presidency.

During that time, the United States started to surpass the Soviets, slowly but surely. NASA completed the Mercury program with three successful launches and followed with the GEMINI program. All of the Gemini missions took place in Earth orbit, but they all pointed in one direction—the moon.

From March of 1965 to November of 1966, the United States launched an unprecedented ten *Gemini* flights. *Gemini* was a new spacecraft that carried two astronauts into orbit (the program was named after the constellation Gemini, "the Twins.")

Now the United States started setting records for longest space walks, longest missions, and most orbits. Strangely enough, the Soviet Union made no manned launches during this period. In 1967, after the Gemini program concluded, the Soviets achieved another "first" in space, but it was an accomplishment they did not want. The first test of their new *SOYUZ* spacecraft was a failure, as the craft crashed on reentry. Cosmonaut Vladimir Komarov became the first space flier to die on a mission. The Soviets did not fly again for over a year, with a successful flight of another *Soyuz* vehicle in late 1968.

The race was almost over. The United States, riding the momentum of the highly successful Gemini program, began Apollo in early 1967. The American program suffered a severe setback when a fire on the launchpad killed Virgil I. "Gus" Grissom, a Mercury and Gemini veteran; Edward H. White, the first American to walk in space; and Roger B. Chaffee, who had not previously flown.

The *Apollo 1* accident triggered an investigation and major modifications of the Apollo program, but did not reduce the nation's commitment to the moon program. Like the Russians, the Americans did not fly again for some time, but when they did with *Apollo 8*, it was with daring and skill that could not be matched.

The astronauts on the *Apollo 8* mission did not land on the moon. That would come later, with *Apollo 11*. However, *Apollo 8* was almost as important and dramatic as the historic lunar landing mission.

Nineteen sixty-eight had been a terrible year for the United States in terms of the politics of Earth. The United States had become deeply engaged in the war in Vietnam under President Lyndon B. Johnson. However, the war had grown increasingly unpopular at home, especially among students.

On March 31, President Johnson surprised the nation by announcing that he would not run for reelection as president in the fall of 1968. He halted the American bombing of North Vietnam and offered to enter into peace talks with the North Vietnamese.

Only five days later, civil rights leader Martin Luther King was assassinated in Memphis, triggering riots in cities across the nation. A few months after that, Senator Robert F. Kennedy was also struck down by an assassin's bullet, just after winning the California presidential primary. American felt that they had lost a war and three of their leaders in five short months. Where could they turn to find renewed hope?

For some, *Apollo 8* provided it. The mission showed once again that the United States could accomplish great tasks, even during very difficult times. In the long history of human exploration, the *Apollo 8* mission set a high standard for courage and risk-taking. At the time, the giant *Saturn V* rocket had never been used for a manned flight, and *Apollo 8* had originally been planned as a testing flight in Earth orbit. At the last minute, officials decided to go all the way to the moon, orbit it, and return. Why?

At the time, the Soviets were sending unmanned spacecraft to the moon. It appeared that they might be planning a manned landing soon. Perhaps it is for this reason that the United States made the bold

Figure 15—7:51 a.m., December 21, 1968. The Apollo 8 _space vehicle lifts off from the launchpad at Kennedy Space Center in Florida. This historic flight would be the first to orbit the moon._

decision to "go for broke" with *Apollo 8*. Perhaps the need to provide the American people with an uplifting conclusion to a difficult year played into that final decision as well.

Whatever the reason, the biggest rocket ever built up until that time roared off the launching pad at Cape Canaveral on December 21, 1968, carrying three astronauts, Frank Borman, James A. Lovell, Jr., and William A. Anders, toward the moon. It had been only forty-two years since Goddard had built the Saturn's ancestor in western Massachusetts.

It was also only twenty years since the astronomer Fred Hoyle had predicted that seeing the Earth from space would create a new revolution in human thinking. That revolution began with *Apollo 8*.

The major purpose of the mission was to go to the moon, and everyone was anxious to see the Earth's satellite close up. However, once the trip was under way, it became clear that seeing the Earth from space was just as important. In all the previous flights, astronauts had seen the Earth from orbit, just a few hundred miles from the surface. Now, at one hundred thousand and two hundred thousand miles, they could see it completely, as a little blue sphere against the black background of space.

For the first time in history, humans saw the whole Earth from space. They saw how small, fragile, and beautiful it is. They felt how serene and lovely it is, even when people are fighting among themselves on its surface. Someone called it "the big blue marble," a name that has stuck ever since.

The astronauts reached the moon on Christmas Eve and went into an orbit that took them as close as seventy miles from the lunar surface. The lifeless

Figure 16—An "Earth rise" viewed over the moon's dark, uneven surface. Astronauts have often commented on the contrast in space between the beauty of Earth and the "vast, forbidding" surface of the moon.

moon impressed the astronauts because it was so different from the life-giving Earth. One of the first descriptions of the moon from close up was that of Frank Borman, who called it, "vast, lonely, and forbidding."

As the astronauts orbited the moon for the first time, they were treated to another new sight—Earthrise. The Earth hung before them in the blackness of space, the top half a beautiful blue and white, the bottom half in darkness. A half-moon was familiar to the human eye, but this was the first half-Earth ever! Seen from the *Apollo 8* spacecraft, the Earth and the moon had traded places. As with *Sputnik*, people could never think about the Earth, or space, in quite the same way again.

On Earth, it was easy to see the differences dividing humans into warring camps. From the moon, the astronauts saw what united us, our common home on a planet without borders or boundaries.

Apollo 8 raised the spirits of Americans and of people everywhere on the planet. It was an American mission, but it helped Americans and the citizens of other nations to feel the unity of all those on the "good Earth."

Apollo 8 also meant the beginning of the end of the space race as it had been viewed up until that time. *Apollo 9* and *10* provided further testing of the systems that would take men to the moon and land on its surface. *Apollo 10* returned to the moon, and even sent the lunar lander to within just nine miles of the moon's surface.

The Soviet Union continued to make flights to Earth orbit, but by mid-1969, it seemed that the United States had the moon to itself. The Soviets simply could not match the *Saturn V*, built by Von Braun and his team. All that remained was for astronauts Neil Armstrong, Edwin "Buzz" Aldrin, and Michael Collins to fully realize President Kennedy's dream.

On July 16, the *Saturn V* once again lifted off from Cape Kennedy (Cape Canaveral was renamed Cape Kennedy for a period after the president's assassination). This time, the purpose of the mission was well

Figure 17—July 16, 1969. The 363-foot-tall Apollo 11 *space vehicle is launched from Kennedy Space Center in Florida on its historic journey to the moon's surface, manned by astronauts Collins, Armstrong, and Aldrin.*

known to Americans and the world. Unlike *Apollo 8*, everyone knew what *Apollo 11* was meant to do, and thousands of dignitaries, reporters, and ordinary people flocked to "the Cape" to see the launch.

Two of the men riding on top of this rocket would be the first humans to walk on soil not of this Earth. They carried with them the spirit of thousands of explorers who had gone before them. Their feat is certain to inspire the thousands who will come after them.

The voyage out was comparatively uneventful. Even after only two flights, going to the moon had become a bit "ordinary." The real drama began on July 20, when Armstrong and Aldrin separated their LUNAR MODULE (named Eagle) from the COMMAND AND SERVICE MODULE (named Columbia) and began the short flight to the lunar surface. They headed for a spot known as the SEA OF TRANQUILITY.

Armstrong had to take control from the computers at the last minute to avoid crashing the Lunar Module in a boulder-strewn crater. The people in Mission Control, 240,000 miles away in Houston, Texas, listened, afraid to speak or breathe. Armstrong and Aldrin described the landing as they approached the surface . . . then there was silence. After a moment, Mission Control heard the first words spoken on the moon. The old pattern was finally reversed—the words were in English, not Russian, and the first word was the name of an American city.

"Houston," said Neil Armstrong, "Tranquility Base here. The Eagle has landed." The first reply to a person on the moon reflected the relief that was felt back on Earth. Astronaut Charles Duke said, "Roger,

Figure 18—A view of the Command and Service Module (CSM) from the Lunar Module (LM) during the Apollo 15 *mission in July 1971.*

Tranquility, we copy you on the ground. You've got a bunch of guys about to turn blue. We're breathing again. Thanks a lot."

One more moment of great drama remained: Neil Armstrong, commander of the mission, would also be the first human being to set foot on the moon. The world waited to hear what he would say, since his statement had not been released to the press.

After several hours of preparation, Armstrong

donned his space suit, made his way cautiously down the ladder of the Lunar Module, and stepped onto the moon. As he did, he said, "That's one small step for man, one giant leap for mankind." (In the years since, his statement has been changed to read, "one small step for *a* man," since "man" and "mankind" really mean the same thing.)

Armstrong was followed onto the surface by Aldrin. Together, they spent a total of two hours and

Figure 19—Astronaut Edwin E. "Buzz" Aldrin, Jr. takes a sample of lunar soil during the Apollo 11 *mission in July 1969. This photograph was taken by Mission Commander Neil A. Armstrong, the first American to walk on the moon.*

twenty-one minutes on the moon. They collected moon rocks, raised an American flag, and left a plague that read: "Here men from the planet Earth first set foot upon the moon July 1969, AD. We came in peace for all mankind."

After a total stay on the moon of just under a day, Armstrong and Aldrin fired the Lunar Module's AS-CENT ENGINE. They headed toward the Command Module's pilot, Michael Collins, who must have been the loneliest man in the universe at that time. Because of the last-minute change in landing sites, Collins had not known where "Tranquility Base" was.

It was only when he actually saw the landing craft rising up from the surface that he knew the location of Armstrong and Aldrin. After a successful docking of the two spacecraft, *Apollo 11* headed back to Earth. The Apollo program had won the space race.

How did the world react to the incredible success of the moon landing? President Richard Nixon made a telephone call from the White House to the astronauts, which was relayed by radio to the moon. The president caught the spirit of the moment when he told them: "For one priceless moment in the whole history of man, all the people of this Earth are truly one. . . ."

For many, *Apollo 11* was overwhelming and difficult to grasp. You might think that there would have been nothing but great joy and excitement everywhere on the planet. There was much of that. However, many people felt uncertain about the value of the moon landing. The war in Vietnam continued, racial problems persisted, and the United States remained deeply divided at home.

When the *New York Times* interviewed a range of prominent people for their reaction to the moon mission on July 21, 1969, they got a variety of answers.

The Reverend Jesse Jackson, a former aide to Martin Luther King, and future candidate for president of the United States, said:

> It seems to me that only with mixed feelings and considerable misgivings can the person of moral concern contemplate America's moon shot and man's first extraplanetary steps. . . . While we can send men to the moon . . . we can't get foodstuffs across town to starving folks in teeming ghettoes.

The Dalai Lama, spiritual leader of Tibetan Buddhists, looked forward to meetings with advanced beings from other planets, and said:

> . . . the most wondrous event would be if men could relinquish all the stains and defilements of the untamed mind and progress toward achieving the real mental peace when he reaches the moon.

Perhaps it was R. Buckminster Fuller, architect, poet, and philosopher, who coined the term, "Spaceship Earth," who most fully understood the landing. Urging the continuation of a strong space program, Fuller said:

> Not only must we put our space programs on highest priority of attention . . . but all humanity must enter into a new reeducational system geared to develop our awareness that we are indeed in space.

The space race, as defined by President Kennedy, was over. If we had learned anything from the early national space programs, it was that we are in space. The Earth, as Copernicus had said, is a planet traveling through space. Buckminster Fuller called it "Spaceship Earth," and space exploration continued to cause people to think differently about space and Earth. From a totally competitive attitude in the early 1960s, the nations moved to a tentative understanding that we had to go into space "in peace for all mankind."

BACK TO BASICS (1970-1985)

Most people do not know that Christopher Columbus made several trips back to the "New World" after his historic landing in 1492. Most people also do not realize that the Apollo program sent six more missions to the moon after *Apollo 11*. All of them were successful except *Apollo 13*, which was cut short because of an explosion on board the spacecraft on the way to the moon.

Apollo continued until December 7, 1972. Almost four years to the day after *Apollo 8* first orbited Earth's nearest neighbor, the last man walked on the moon. Since that time, no one has gone to the moon, and the only *Saturn V* rockets in existence are "museum pieces."

Apollo was a great journey, but it was also something of a detour. From the earliest days of planning manned space exploration, it had been assumed that after the rockets were developed and the first satellites launched, the next step would be a permanent presence in Earth orbit—a space station. From there, astronauts and scientists could make useful observations of the Earth. In addition, space stations would

Figure 20—A view of scientist-astronaut Harrison H. Schmitt on the moon's surface during the last moon mission in December 1972.

be jumping-off points for missions to the moon, and Mars and other planets.

With Apollo, the United States skipped what many had thought would be a major step in space, the permanent space station. However, the Soviet Union seemed to take that approach quite seriously.

Once it became clear that the United States would win the race to the moon, the Soviets began to establish an impressive record in Earth orbit.

Throughout the 1970s and 1980s, the Russian cosmonauts steadily built up their record-breaking "stays" in orbit: 63 days in 1975; 96 days in 1977; 139 days in 1978; 175 days in 1979; 184 days in 1980; 211 days in 1982; 236 days in 1984. In 1987, a Russian space flier came close to a year in space. In 1988, two cosmonauts spent just over a year in orbit.

The Soviets did not go to the moon, but they did achieve another major "first" in space: by 1988, their *MIR* ("Peace") station was indeed permanently occupied by rotating crews of cosmonauts.

The United States also built a space station just after the Apollo missions ended. SKYLAB was placed in orbit in 1973, and remained there until it slipped into the atmosphere and burned up in 1979. Three crews manned *Skylab* on three separate occasions until early 1974. The last crew briefly held the record for time in orbit.

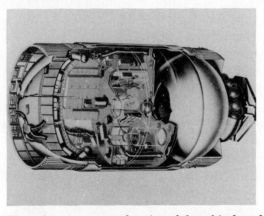

Figure 21—A cutaway drawing of the orbital workshop of the Skylab space station. Skylab weighed over 100 tons and was manned for several months during 1973 and 1974 by NASA astronauts.

In the early 1970s, the United States even worked on a comprehensive plan for following up the Apollo missions. A task force chaired by Vice President Spiro Agnew produced a report calling for a vigorous program of exploring the space frontier. However, there was little or no support for major space missions at the time. To many Americans, Apollo had achieved the main goal in space—beating the Russians to the moon. In addition, American society was still distracted by its problems on Earth, including a growing scandal known as "Watergate."

The Nixon administration and Congress responded to the space exploration community's plans by giving NASA just enough money to build a shuttle, and no more. Soon, both President Nixon and Vice President Agnew were mired in the Watergate scandal, which led to their resignations. Without strong leadership from the executive branch, bold initiatives in space simply could not generate enough public or Congressional support. The glory days of Apollo were soon a fading memory.

However, manned spaceflight of the Mercury/Gemini/Apollo type has never been the only approach to exploring space. During the period from 1970 to 1985, there were many new developments in the unmanned area that caused people to think in new ways about space.

Astronomers continued to study space from Earth. They also had their first opportunity to study the universe from a position in space itself. This allowed them to see much more, because the Earth's atmosphere no longer obscured their vision. Using a telescope launched by the space shuttle, astronomers made a number of important discoveries, including strong evidence of planets revolving around a distant star. Until that time, astronomers had assumed that other stars had planets, but there was little to confirm it.

During this period, the Search for Extraterrestrial Intelligence (SETI) picked up momentum as well. SETI is based on the idea that if intelligent life has evolved on other planets, they may be trying to signal us. Using powerful "radio telescopes" (systems that listen for radio waves, rather than picking up the light of stars), SETI experts began to sweep the sky for signals that might be sent by beings on other planets. So far, they have heard nothing to convince them of an extraterrestrial presence.

The United States and the Soviet Union flew their first (and only) cooperative manned flight in 1975. The APOLLO/SOYUZ mission involved the launching of separate spacecraft from the United States and the Soviet Union. They then docked in orbit and flew together for several days. This was officially the last Apollo flight (Apollo 18).

While *Apollo/Soyuz* orbited the Earth, the sixth anniversary of Apollo 11 was observed. What a difference in only a few years—from a highly competitive "space race" to the first cooperative venture of the two leading spacefaring countries. Again, activities in space reflected the politics of Earth. The United States and the Soviet Union had entered a period of détente (cooperation) under President Nixon and Premier Brezhnev; the *Apollo/Soyuz* mission was meant to symbolize the new relationship.

The United States did not have any manned flights after *Apollo/Soyuz* in the 1970s. Instead, much of NASA's energy went into the building of the space shuttle. However, the unmanned program remained active and successful. The MARINER and VIKING

Figure 22—Scientists engaged in SETI research listen for signs of intelligent life in outer space. So far they have heard nothing certain but they remain hopeful.

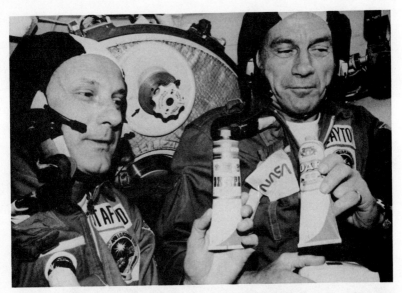

Figure 23—NASA astronauts Thomas P. Stafford, left, and Donald K. "Deke" Slayton visit with Soviet cosmonauts during the Apollo-Soyuz _rendezvous in 1975. They are raising tubes of Russian borscht (beet soup) that have been pasted over with Russian vodka labels in a "toast" to the cosmonauts._

programs sent spacecraft to Mars, with the first landing on that planet in July 1976. The _Viking_ lander even tried to determine whether there was life on Mars, but the results were unclear.

The PIONEER and VOYAGER spacecraft toured much of the solar system, providing invaluable new knowledge of the outer planets and their moons. The Pioneers became the first objects made by humans to leave the solar system altogether.

Finally, on April 12, 1981, the twentieth anniversary of Yuri Gagarin's flight, a new kind of spacecraft left the Earth on its first voyage. After years of effort, the United States had built and launched the first

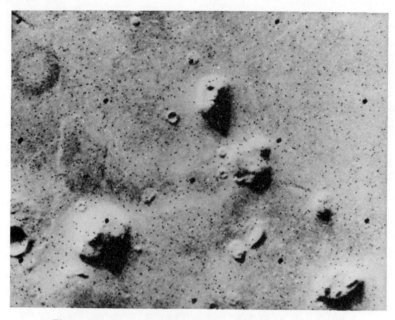

Figure 24—A photograph taken by Viking I *of the surface of Mars. The terrain is similar to the deserts of the American Southwest. The rock formation at the top resembles a human face with shadows giving the appearance of eyes, nose, and mouth.*

reusable spaceship, the SPACE SHUTTLE. The shuttle flew successfully until early 1986, though its twenty-four flights in five years never approached the original goal of two per month.

The shuttle had been sold to Congress and the American people as a "space truck" that would bring routine access to space. The shuttle has not done that. However, it has helped people to think about going into space as routine. Several nonastronauts flew on the shuttle as "payload specialists," including a senator and a congressman. NASA also inaugurated the "Teacher-in-Space" and "Journalist-in-Space"

Figure 25—The launch of the first space shuttle, Columbia, *in April 1981. The shuttle was designed to serve as a "space truck," bringing people and material in and out of orbit with relative ease.*

programs, which were cut short by the Challenger disaster in 1986.

The idea of routinely going into space also helped people to consider their own roles in space exploration differently. In the 1960s, there had been only two nations in space, the United States and the Soviet Union. In both countries, government programs dominated space exploration activities, with little or no private efforts. Only a few highly trained individuals could imagine becoming astronauts or cosmonauts.

From 1970 to 1985, the picture changed dramatically. During that time, there were major changes in the space field:

- Many new national space programs began;
- Commercial opportunities in space started to look promising;
- Private space-oriented groups appeared on the scene.

In addition, President Reagan inaugurated a major shift in public understanding of space in 1983, when he announced his STRATEGIC DEFENSE INITIATIVE (SDI), popularly known as "Star Wars." His concept of placing a "defensive shield" against nuclear weapons in space once again raised the issue of how space would be used for military purposes.

Let's look at each of these changes in turn.

First, new *national space programs* appeared during this period, and organizations such as the EUROPEAN SPACE AGENCY (ESA) emerged as major forces. ESA includes twelve member nations and one associate. The agency develops projects that are too expensive for any one nation to undertake alone.

The *Ariane* rocket is an example of this approach. It is an unmanned vehicle designed to put satellites into orbit. *Ariane* effectively competes with the American space shuttle and other launch vehicles. ESA is also participating in the American-sponsored international space station, *Freedom*.

By the mid-1980s, ESA members had been joined by countries such as Japan, China, Sweden, Brazil, India, and Australia as nations with a stake in space exploration and development. Japan and China, for example, have their own satellite-launching capabilities. Australia announced plans for development of the first commercial spaceport in the world. India is using satellites to help educate its enormous population.

However, governments are no longer the only players in the space game. The first successful *commercial development* of the space environment followed directly behind the first activity in space. *Sputnik I* pointed to a future when there would be many communications satellites in orbit. By the mid-1980s, the communications satellite business was a mature industry worth billions of dollars.

Satellites have also become critical to weather forecasting, and there is hope that "remote sensing" will be another commercial use of satellites. REMOTE SENSING SATELLITES are used for purposes such as detecting resources or monitoring crops. Companies can make money from them by putting them into orbit, or selling the pictures they take.

Satellites can also be used as part of a location-finding system. Connected with computers on Earth, they can help to find airplanes that have crashed, locate lost hikers, or track huge fleets of trucks.

CASE STUDY

The Strategic Defense Initiative (SDI)

In early 1983, President Ronald Reagan dropped a bombshell into the debate over the appropriate uses of outer space. He announced that he would commit the United States to the development of a "space shield" that would render nuclear weapons obsolete. Immediately dubbed "Star Wars" by the press, after the popular movie of the same name, the Strategic Defense Initiative (SDI) once again raises the question of the proper military role in space.

The president's plan called for research and development of weapons that could be put into orbit around the Earth. These weapons would be able to detect the launching of missiles from the Soviet Union, or other countries, and shoot them down before they reached the United States.

President Reagan assured the world that his Strategic Defense Initiative was just what he called it—defensive in nature, and posing no threat to other nations. Opponents of the plan, in the United States and elsewhere, responded that "defense is in the mind of the beholder." How, it was asked, could the Soviets, for example, be sure that the advanced high-technology laser beams mounted on orbital battle platforms would not be used to attack Soviet cities?

President Reagan, to the astonishment of friend and foe alike, responded that the United States might even share SDI technology with the Soviets. Why not? he said. That would prove the peaceful intentions of the Americans.

Former vice president Walter Mondale tried to make SDI an issue in the 1984 election, but he failed to make an impression. President Reagan later argued that it was the "Star Wars" plan that helped bring the Russians to the bargaining table and produced a treaty to ban intermediate nuclear weapons in 1987. His critics disagreed and continued to argue that if SDI were ever deployed, it would have a terrible impact on prospects for world peace and would "weaponize" space in a totally unacceptable way.

As President Reagan's term drew to a close, research for his initiative continued to be funded, but actual deployment seemed a long way off. Once again, however, the ongoing debate over civilian vs. military uses of space had been ignited and continues to smolder even today.

Many companies are also conducting research in the area of *materials processing* in the weightless environment of space. The pull of gravity distorts the manufacturing of many objects on Earth. In ZERO-G, materials can be made that are purer or stronger. Crystals developed in microgravity might have applications in developing drugs to cure or treat many common diseases.

Another change in thinking about space evolved in the 1970s and early 1980s with the appearance of *private organizations* interested in many aspects of space exploration, such as space settlement.

A typical organization of this type was born about the time of the *Apollo 11* landing in 1969. Dr. Gerard K. O'Neill, a professor at PrincetonUniversity, asked one of his classes to think beyond Apollo. He challenged them to consider how large numbers of people could live in space. The class discussion led to a revolutionary idea: instead of living on a planetary

Figure 26—An artist's rendering of the LANDSAT satellite.
LANDSAT *was the first satellite devoted solely to the study of Earth's resources from space. Satellites have helped humanity to communicate more effectively and will soon help us to better manage our planet's limited resources.*

Figure 27—An artist's rendering of a chemical processing plant in space. In the not-too-distant future, plants of this type may provide the Earth with cheaper and better materials.

surface, such as the moon or Mars, humans could live in "space habitats." These settlements, built out of extraterrestrial materials brought from the moon or the asteroid belt between Mars and Jupiter might house up to 10,000 people!

The space settlements, it was found, would best be placed at the "libration points," or LAGRANGE POINTS between the Earth and the moon. At these points, the pull of gravity from the Earth and the moon would be balanced, providing the space settlements with maximum stability. The best Lagrange point, named for the French scientist credited with their discovery, was thought to be the fifth one, "L5."

This way of thinking about space was a real breakthrough. Until that time, few had imagined that large numbers of people could live comfortably in artificial environments in space. O'Neill's original work is carried on by the Space Studies Institute (SSI), founded in 1975.

The L-5 Society grew out of the work of the Space Studies Institute. A grassroots, "prospace" membership organization, L-5 envisioned holding its last meeting at the fifth Lagrange Point!

Wernher Von Braun founded another private space-oriented organization, the Washington-based National Space Institute, in the 1970s to help keep up public interest in a strong space program.

L-5 and the National Space Institute merged to form the National Space Society (NSS) in the mid-1980s. NSS is one of the most active organizations in the United States advocating space exploration and development. Its stated mission is the creation of a spacefaring civilization that will establish communities beyond Earth.

Many other organizations were founded after Apollo and before the *Challenger* accident. These have included the Planetary Society, U.S. Space Foundation, and Institute for Security and Cooperation in Outer Space (ISCOS).

Young people have been active in founding and running the Students for the Exploration and Development of Space (SEDS), which focuses on space-related issues at the college level. Space Camp, an idea first advanced by Wernher Von Braun, began in Huntsville, Alabama, in 1982 with 700 participants. By 1988, Space Camp had expanded to other locations, added a Space Academy and an adult program, and counted 18,000 participants.

Why did all of these new groups appear after the

space race apparently had ended? Does it mean that there is something exciting about space exploration other than the "race"? Does it mean that many more people want to be involved in space exploration and development, and that if they cannot get into space through the existing national space programs, they will find a way to go there themselves? What do you think?

RETHINKING SPACE EXPLORATION (1986–1988)

Florida is known for its tropical climate, but it was very cold at the Cape the morning of January 28, 1986. Ice hung from the girders of the launchpad. The space shuttle *Challenger*, which had made several successful flights into Earth orbit, stood patiently waiting to fly again.

The launch had been delayed several times. Many of the dignitaries who had flown to Florida for the occasion were back in Washington, getting ready to hear the president's State of the Union address.

The nation's unmanned space program had scored a triumph over the weekend, as the *Voyager* spacecraft sent back beautiful television pictures of the planet Uranus and its moons. But the shuttle still sat and waited.

The pressure mounted to launch the *Challenger*. This was supposed to herald a new era in space, a time when ordinary citizens would fly. Christa McAuliffe, an exuberant and inspiring teacher from Concord, New Hampshire, would be aboard. She had won NASA's "teacher-in-space" competition, and she was going to conduct the first class from orbit.

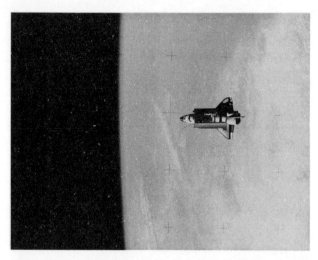

Figure 28—A view of the space shuttle Challenger *orbiting Earth in June 1983. Thirty months later, in January 1986,* Challenger *would explode, killing its seven crew members. The* Challenger *disaster forced an overhaul of the shuttle's design and a reassessment of its role.*

Students all across the country gathered around television sets to watch the launch.

Six others were aboard that day, one woman and five men. Five were career astronauts; one was a payload specialist riding along with his company's satellite.

NASA managers and engineers who had helped design the shuttle conducted intense debates into the night. Finally, the decision was made to "go," in spite of the cold weather. The countdown commenced, and the shuttle lifted off. For the first minute or so, all seemed to go well. Then, as Mission Control called for "throttle up," a heart-stopping explosion separated the shuttle orbiter from the external tank, and the two solid fuel rockets headed into the sky, out of control.

For the first time in the history of the American space program, astronauts had been killed while on a mission.

Later, an investigation by a special committee showed that the accident was caused by the failure of an O-RING. This was a part that some people at NASA had suspected could be dangerous. The investigation also showed that the engineers from the company that made the solid rocket boosters had opposed the launch. The shuttle had never been launched in such cold weather, and they did not think it was safe. The commission found that NASA had been pushing the shuttle beyond its capabilities. NASA had promised to launch twice a month, and it was trying to keep its promise. But the shuttle proved to be too delicate and complex to meet that schedule.

The *Challenger* disaster forced a radical rethinking about space in America. It also changed the direction of space exploration in specific ways. The disaster:

1. Showed that access to space had not become routine, and put NASA's "Citizen-in-Space" programs on indefinite hold;
2. Increased pressure on NASA to define its long-range goals in space;
3. Helped to force a new national space policy into existence;
4. Created many new commercial opportunities in space.

NASA backed away from the argument that access to space had become routine and embraced the demand for long-range goals. The president's NATIONAL COMMISSION ON SPACE had begun work before the *Challenger* disaster, and it issued its report in the spring of 1986. The report offered a new series of long-term goals for NASA and the nation. It stated that "the solar system is the extended home of humanity." It called for a visionary fifty-year plan to "pioneer the space frontier," culminating with a human outpost on Mars.

NASA accepted many of the commission's recommendations. The agency also assigned astronaut Sally Ride to conduct a study of appropriate goals. Her report suggested a "mission to planet Earth." This idea envisions the use of space technology, such as remote sensing satellites, to help solve problems on Earth by finding resources, predicting droughts, etc. Dr. Ride's report also outlined approaches to exploration of the solar system, a return to the moon, and a mission to Mars.

The reports of the National Commission on Space and Dr. Ride's recommendations formed the basis of a new national space policy, issued in early 1988. The new policy called, for the first time, for an expanded human presence in the solar system. It also increased support for private/commercial activities and reduced the shuttle's dominant role in American launch capabilities.

President Reagan took the shuttle out of the commercial satellite-launching business even before the new policy was issued. This opened up many opportunities for foreign companies and manufacturers of unmanned launch vehicles. With the shuttle out of commission, new relationships developed internationally. For example, an American company, Payload Systems, Inc., made an agreement with the Russians to conduct a series of zero-gravity experiments on the Soviet *MIR* space station. The Chinese visited the United States, trying to market their "Long March" rocket to launch American satellites.

For years, the idea of a space race had not been an issue. As with *Sputnik*, however, the image of the race emerged again when the American space program faced a serious crisis. In the wake of the *Challenger* accident, observers recalled the race between the tortoise and the hare. The Soviet Union looked like the tortoise—slowly and carefully plodding along, but always making progress. The United States looked like the hare—racing far ahead, and then getting distracted and falling by the wayside.

The American space program did recover from the *Challenger* accident. It took intense analysis, redesign, and testing, but on September 29, 1988, the space shuttle *Discovery* lifted off from Cape Canaveral and began a near-flawless mission. The *Discovery* astronauts launched a replacement for a satellite that had been lost on the *Challenger* mission, and read tributes to those who had died in the disaster. For the first time in over two and a half years, those managing the space program felt they could look ahead instead of back.

There was one other important rethinking that took place after the *Challenger* accident. The disaster reaffirmed the belief of many who wanted to get into space that they could not count on the American space program alone to get them there. Like their Puritan and Pilgrim forefathers who struggled to find their own way to the "New World," these latter-day pioneers now knew that they would have to do it themselves.

REVIEW QUESTIONS

1. Why did the United States react so strongly to the launching of *Sputnik I?*
2. Was there really a "space race"? Do you think that competition is the right way to explore and develop outer space?
3. How has the exploration of space reflected the politics of Earth?
4. Which moon mission do you think was more important: *Apollo 8* or *Apollo 11?* Why?
5. Why didn't everyone react to the moon landing with support and excitement?
6. Compare the American and Soviet space programs in the 1970s and 1980s. Who had the better strategy?
7. Why did the *Challenger* disaster force a radical rethinking of American space policy?
8. How is the space exploration field different in 1988 as compared to 1968, when *Apollo 8* went to the moon?

4 | The Issues of Today

In late 1968, only two nations of Earth had sent their citizens into outer space. The issues of the day seemed simple, and focused on which of them would win the so-called space race.

Today, only twenty years later, the picture has changed dramatically. Instead of two spacefaring nations, there are as many as forty countries with some involvement in space exploration. Instead of one overriding issue, there are a multitude.

In this chapter, let's consider how people are thinking about space today. What are a few of the issues that must be decided in the next decade?

EARTH vs. SPACE

One issue has remained the same since the beginning of the Space Age. It is the idea that we should not spend money on space when we have so many problems on Earth.

Remember what Jesse Jackson said about the moon landing? How can we be totally happy about landing on the moon when we can't get food to the ghettos? Twenty years later, not much has changed. The United States still has a space program, as do a number of other nations. The United States still has poverty and hunger, as do most nations on Earth.

Those who oppose massive expenditures on space exploration argue that the money should go into feeding the hungry, housing the homeless, or providing jobs for the unemployed. On a global scale, countries in the Third World fear that the more developed nations will increase their wealth in space, leaving them in a "planetary ghetto" on Earth. Few of them would argue for elimination of space exploration altogether, but they would put a priority on solving Earth's problems first.

Supporters of space exploration respond that only a dynamic society can generate the wealth and the will to solve social problems. They argue that space exploration contributes to that sense of dynamism

Figure 29—A homeless man rests under the cover of a newspaper bearing news of the Apollo 11 *moon landing. Many Americans feel that the money spent on space exploration would be better used on social programs.*

and purpose. Many suggest that space exploration can help to solve problems on Earth, as with the "Mission to Planet Earth." In reference to the United States, they also point out that the nine billion dollars spent per year on the civilian space program is small compared to the defense budget and other government programs.

Of course, the best approach would be a global approach both to eliminating hunger and poverty and exploring space at the same time. But right now, that isn't happening. In the meantime, what do you think we should do, as a country and as a planet? Should we spend more on space exploration or less? Should we wait until more of our problems on Earth have been solved, or go forward with space exploration no matter what?

WAR vs. PEACE

Those who are concerned about rapid development of the space environment also worry about the military's role. Many feel that the military will, or already does, dominate the American and Soviet space programs. When they think of space, they think of "Star Wars" (SDI) and they do not want to see the arms race extended off the planet. In the United States, the civilian and military space programs were separate for a long time. Now, however, the shuttle is used for both civilian and military missions. The Soviet Union has never made a distinction between military and civilian operations. The European Space Agency (ESA) does have a clear policy on this matter: no military activities.

Many people in the United States feel that, whether

Figure 30—What, if any, role should the military have in space exploration? This is one of the most hotly debated questions that we face as we think about our future in space.

we like it or not, we must have a strong military presence in space to counter the Soviet Union (there are probably people in the Soviet Union who feel the same way about us). They see the extension of the competition between different "ways of life" as inevitable.

Others argue that we have an extraordinary opportunity to start afresh on the space frontier, putting aside the rivalries that have led us into wars on Earth. Supporters of a strong civilian space program often share these concerns about military involvement.

Many of them oppose the "Star Wars," or Strategic Defense Initiative program.

However, it is hard to see how the military can simply be excluded from the space environment. One approach, advocated by the Institute for Security and Cooperation in Outer Space (ISCOS) and others, is to accept certain types of a limited military presence in space, but oppose "weaponization." For example, military satellites can be useful in keeping the peace, by monitoring arms control agreements. Military personnel can help in settling difficult environments, such as the moon. Many advocates of peaceful space exploration would accept these military uses of space. However, they would oppose putting massive weapons systems, such as laser battle platforms, in orbit.

If you were the president of the United States, and you had to take a position on this issue, what would it be? Do you think a strong military presence in space is necessary? Would you limit the military role in the space environment, or allow the military to extend its reach to the moon and Mars? As a citizen, do you feel that the importance of space exploration outweighs the risks involved with a military presence in space? Or would you prefer to shut down all space-related activities if the military is going to be a part of them?

COOPERATION vs. COMPETITION

As we have seen, competition dominated everyone's thinking about space throughout the 1960s. Many still see it that way. However, there are many more opportunities for international cooperation in space, and many examples of its taking place. The United States is cooperating with Japan, Canada, and the

European Space Agency in building its space station, for example.

However, several government agencies have blocked American companies' use of Soviet rockets for launching purposes. Except for these examples, international cooperation is extensive in the commercial area. Companies are simply picking the lowest cost approach to achieving their goals.

Many in the American government also oppose a joint United States/Soviet Mars mission. The Soviet Union favors such as mission, as do many American space activists.

The Soviet Union has also proposed formation of a "World Space Organization" to coordinate cooperative ventures on the final frontier. So far, the United States has consistently opposed this concept.

The issue of cooperation between developed and undeveloped nations remains unresolved as well.

In its report, the president's National Commission

Figure 31—A Soviet cosmonaut and an American astronaut greet each other in space. Whether or not the superpowers can cooperate will determine whether we have a competitive, and possibly violent, future in space or a cooperative and productive one.

on Space said this about cooperation: "Exploring the universe is neither one nation's issue, nor relevant only to our time. Accordingly, America must work with other nations in a manner consistent with our Constitution, national security, and international agreements."

Do you think that the United States (and other nations) should focus more on cooperation or competition in the exploration and development of space? If the wealthier nations do most of the investing in space projects, do they have an obligation to share the results with less developed countries? Should the United States cooperate in space with nations that compete with it on Earth?

PUBLIC vs. PRIVATE

Who owns outer space? Have you ever wondered whether you could build a house on the moon and live there the way you do on Earth? Or could the United States have laid claim to the moon as its territory when the Apollo astronauts went there?

This is another issue that must be resolved before the space frontier can be developed.

Some argue that if outer space belongs to anyone, it belongs to all of humanity. If that is so, then no one should be allowed to claim or own a part of the moon, or Mars. Also, whatever is done in space should be for the good of all humanity. No one country should be able to exploit the space environment for its own gain.

Those in favor of private enterprise in space respond by saying that there will be no space development if people cannot profit from it. Why would people go to live in a space settlement if they couldn't improve their lives and their families' lives? Why would someone invest in a space-oriented business if he or she couldn't make money from it?

Currently, space law leans toward letting people make money in space (commercial satellites are an example) without allowing them to own property there. You, the reader of this book, may be called upon to vote on this issue sometime in the future. How will you decide? Should everyone share what is available in outer space, or should some nations and individuals be able to get more than others?

EXPLORATION vs. DEVELOPMENT

Many people are in favor of exploration and development of space. Others feel good about exploring, but are concerned about developing. For example, the idea of a scientific outpost on Mars is exciting and

Figure 32—Should individuals be permitted to own property in space or should space belong to all mankind? This is an important question that we must answer before we expand our presence in space.

invigorating. However, can you imagine a Mc-Donald's on Mars? How would you feel about that?

Just about everyone feels that exploration is a good thing—not just the exploration of space, but all kinds of exploration. Without it, people and societies can become "stuck," mired in the old ways of doing things. However, development is less clearly beneficial. Those who are concerned about space exploration ask, "Won't we just mess up space the way we have messed up the Earth?" Or they say, "Are we mature enough to do better than we have in the past?"

These are difficult questions. Even today, the people of Earth face staggering problems resulting from misuse of technology and rapid development. We are already experiencing "global warming" and losing gases from the atmosphere because we have abused the Earth's environment while exploring it.

Supporters of exploration and development argue that sending people into space will actually help to preserve the Earth. We can get nonpolluting solar energy from space, move factories into orbit, and relieve the population pressures with space settlement. We have learned from our experience on Earth, they say, and we will do better on the moon, on Mars, and in other places in space.

What do you think? Is there any way to assure that humans in outer space will behave better than humans on Earth? Would a Disney World on the moon be a bad thing? Should the solar system be made a wilderness area, like our national parks? These decisions must be made soon, because once human expansion into space begins on a large scale, it will be hard to change directions.

MANNED vs. UNMANNED

For some, especially those in the physical sciences, the answer to many of these questions is simple: keep human beings out of space. They argue that we can learn far more for far less investment through un- manned missions. They point out that a manned mission to Earth orbit, the moon, or Mars requires great expenditures in order to keep the crew alive. They also do not believe that the addition of human crew members adds to the value of the information we get from space.

Their opponents agree that scientific efforts should not necessarily include human crews. However, if one goal of space exploration is for human beings to live and work in space permanently, then we must send out manned missions now. We need to learn how people respond to the space environment phys- ically and psychologically, and we need to know the barriers to successfully functioning there.

Supporters of manned space exploration also sug- gest that support for space exploration will diminish without manned flights. They further argue that it is part of the human destiny to leave planet Earth and expand into the universe. This cannot happen with an unmanned program alone.

Most countries have chosen a middle road in their space programs so far. They support both manned and unmanned missions in space. The idea is being extended beyond short missions as well. After a fierce debate, the United States is now supporting both an unmanned space station developed by a private firm and a permanent manned station to be developed by NASA, ESA, Japan, and Canada.

This debate began when the engineers on the Mercury program tried to put the astronauts into a capsule without windows, and the astronauts demanded not only windows, but also a way to fly the "spacecraft." It is a debate that is sure to go on for quite a long time.

Do you see a way to settle this argument? Should we restrict space to robots and keep humans on Earth, or should we cut down on unmanned flights and put all of our money into space settlements? If we are to take a middle road, how can we define it?

WHAT YOU CAN DO

You can affect all of these debates. In the United States, funding for NASA is an issue that is before Congress in just about every session. The debates over the NASA budget reflect all of the issues discussed here, and you can contact your congressman or senator with your viewpoint on them.

You can also join one or more of the space-oriented organizations listed in the Appendices of this book. All of them are working with their members to move the debates over space exploration in one direction or another.

Or you may be one of those rare people who will lead the space movement in a new direction. There is a lot you can do if you want to be involved.

REVIEW QUESTIONS

1. Should we try to solve all of Earth's problems before exploring outer space? Why or why not?
2. Do you think people should be able to own parts

of the planets in the solar system? Why or why not?

3. Why do some people oppose having the military in space?

4. Is it possible to have space exploration without space development? Can you think of parts of the Earth that have been explored without being developed?

5. Why might the United States refuse to cooperate with the Soviet Union on space exploration?

6. What would be the results of a space program that used no humans at all?

5 | The People of Tomorrow

THE STAR TREK VISION

"These are the voyages of the starship *Enterprise* . . .
Its five-year mission: to explore strange new worlds,
to seek out new life and new civilizations . . . to boldly
go where no man has gone before."

For over two decades, the mission of the *Enterprise*
has fascinated people. It has spawned two television
series, four movies, and a host of related products.
Why is "Star Trek" so popular? Why do some people
seem more interested in it than in the real space
program? Perhaps the answer is that "Star Trek"
offers a vision of space exploration that reflects the
hopes and dreams of those who watch it.

In the "Star Trek" view of the twenty-second century, the Earth is at peace and is also a leader of the
"United Federation of Planets." The crew of the *Enterprise* includes descendants of all the Earth's original races. And they all get along with each other very
well.

For example, the captain and ship's doctor in the
original "Star Trek" are of American ancestry, the
communications officer's background is African, the
helmsman is Asian, and the weapons officer is Russian. In the new "Star Trek," the captain's background is French, the first officer and doctor are

American, and the head of security is apparently Russian.

In both versions, so-called "aliens" play a major role. In the original program, the first officer, Mr. Spock, is descended from a human mother and a father from Vulcan, another member planet of the Federation. In the new program, the ship's counselor is also half-human and half-extraterrestrial. In general, men and women are equals on the starship, and in the new version families are even allowed to be aboard on the voyages.

It may be even more important that the values of the United Federation of Planets are good values, and the people of the *Enterprise* are good people. For example, no starship commander would ever violate the "Prime Directive." The Prime Directive forbids interference in the evolution of any less-developed culture that is encountered on the voyage.

The *Enterprise* explores, but it does not exploit. It seeks knowledge, not wealth. The ship and its crew have weapons, but they are used only as a last resort. All life forms, even those who appear to be enemies, are given respect.

In "Star Trek," the characters do not just explore space. They try to help others, while respecting them as well.

Think about the "Star Trek vision." Does it resolve some of the important questions that were raised in Chapter Three? Does it represent some of your hopes for space exploration?

THINKING BIG

Is the vision presented by "Star Trek" just a fantasy, or could it become a reality? That probably depends on the attitude we take with us as we continue to explore outer space.

Remember the caveman who crossed the icy water hunting caribou? He probably thought of his world as harsh and difficult. There may have been other tribes constantly trying to kill him and his people. He may have had no hope it would get better in the future. But look at what has happened. For all of our problems, most of us would surely say that life has improved since the days of our earliest ancestors. Or think of Christopher Columbus, when he sailed west in search of India. He had big dreams, but wouldn't he be surprised to visit a city like Washington, D.C. (District of Columbia), or to watch the space shuttle *Columbia* lift off from Cape Canaveral? Could he have imagined that there would be places and things like that—named after him?

"Star Trek" imagines the universe as it might look in about two hundred years. Well, it was just about two hundred years ago that the "New World" of North and South America really began. Thomas Jefferson, Benjamin Franklin, George Washington, and the other leaders of that time "thought big." They decided it was time to create a new kind of nation in the New World.

They believed in a nation of people who governed themselves instead of being ruled by a king. At the time, the idea sounded crazy; it was a fantasy. But it did become real. Today, there are many democracies in North and South America. There are also dictatorships, poverty, and many problems. But there are no kings.

Can the people of planet Earth unite and explore outer space as one? Or will the different nations continue to compete for leadership in the solar system and beyond? Can we explore space peacefully, or will the solar system become a new battleground?

Of course, no one can answer those questions for certain. However, each person on Earth can decide to have an impact on the outcome. As we pointed out earlier, the universe is so big that we cannot really imagine its size. It is big enough that just about anything can happen as we begin to explore it. We have only just begun to see what might be possible.

The decisions made by a few people in the eighteenth century have had a big effect on us, their descendants living in the twentieth century. In the same way, the decisions we make today will decide whether the "Star Trek" vision prevails in the twenty-second century.

LIVING IN SPACE

As we have seen, there have been many important moments when people began to think differently about outer space. One of those moments came when Copernicus said the Earth went around the sun. Another occurred when Galileo turned his telescope toward Jupiter and saw its moons. Then there was *Sputnik,* and the flight of Yuri Gagarin, and *Apollo 8* and *Apollo 11.*

The "people of tomorrow" will have their important "moments" as well. One will occur when we really start to live in outer space. Perhaps it will hit home when the first space settlement is established at L-5, or on the moon. Perhaps we will feel it when the first baby is born off the planet Earth. Perhaps you will understand it completely only when some-

Figure 33—An artist's conception of the permanently manned space station. The station will be serviced by the shuttle and heated by the solar panels shown here at the extreme ends of the station. The human inhabitants will occupy the center section.

one you know says he or she is moving—but not to Cleveland, to the moon. The person is going to "migrate" with their family to outer space. Could that someone even be you?

So far, no one who has gone into outer space has intended to stay there. The Russians have had cosmonauts in orbit for over a year at a time, but they have always returned to Earth. Earth is still "home base" for astronomers, SETI investigators, astronauts, cosmonauts, and space entrepreneurs.

During an earlier period of exploration and development on Earth, there was a similar pattern. Those who came to North and South America explored, but did not stay. After a time, the thinking changed, and people came with the intention of remaining in the New World. Even then, they did not cut their ties to the Old World. The Pilgrims, who established a permanent colony in Massachusetts, continued to depend on England for much of their food and other necessities for many years.

In North America, it was the Puritans who began the final break from the mother country. They brought a vision of a new and "purified" society with them to America, and that sustained them during hard times. They also brought enough people, tools, and knowledge to establish self-sufficient communities in the wilderness. Their balance of vision and know-how worked, and their descendants became the founders of a new nation.

It may be a long time before human beings reach the point where everything falls into place and we can support thousands of people living off Earth. However, there are many who want to be "pioneers on the space frontier," and it is likely to happen. When it does, humanity will have to face a number of new questions. For example, what will be the relationships between people on Earth and people in space? If a space settlement wants to declare independence, will Earth allow it?

As humans move into space, we will need not only people who can build space habitats. We may also require a new generation of Thomas Jeffersons, George Washingtons, and Benjamin Franklins. There will be many new challenges on the space frontier.

If new societies are going to develop on this frontier, there must be people of great wisdom and intelligence guiding them.

EVOLVING IN SPACE

The big moment has arrived! The space capsule is floating gently down to Russian soil, bearing its cargo of cosmonauts returning from the space station.

Technicians and members of the support team rush to the capsule and open it. The cosmonauts look out from inside the capsule. They wave to their helpers. Then they climb out and jump up and down to celebrate their return to Earth. Right?

No, wrong! The cosmonauts cannot even climb out, much less jump up and down. More than likely, they are carried out on stretchers and taken to a place where they can rest. After several days, they will be able to walk again.

When people spend long periods in weightlessness, their bodies begin to adapt. Muscles weaken, the bones shed calcium, the flow of blood changes, and the heart beats more slowly.

Think about the body of a fish. It is well adapted for life in an environment similar to Zero-G. When life came out of the oceans millions of years ago, it began to evolve to living on land. We are one of the results of that process.

If people do begin to live in space in large numbers, some will want to have ARTIFICIAL GRAVITY. This is created by putting a "spin" on the habitat in which they are living so that it rotates like a planet. But others may live in Zero-G. As their bodies begin to adapt, they will begin to look very different from humans on Earth. Even on the moon and Mars,

CASE STUDY

George Washington and James T. Kirk

George Washington was an engineer, plantation owner, victorious general, and political leader. A Virginian, he led the colonial army to victory over the British during the Revolutionary War, and was elected the first president of the new republic. He was widely respected and eventually given the title "Father of our Country."

Washington was the kind of multitalented person who flourished in the New World of the eighteenth century on Earth. Rather than being a narrow specialist in one field, he brought to the service of his new country a varied perspective on life. In that way, he was somewhat typical of the leadership of the time. Thomas Jefferson, who also became president, was an accomplished architect and scholar, as well as the political philosopher who wrote the Declaration of Independence. Benjamin Franklin pursued enough careers to satisfy several men, including scientist, diplomat, and printer.

Is there a George Washington in our future on the space frontier? So far, those who have gone into outer space have been specialists trained for specific missions. However, this may change in the future, especially if there is a human population permanently living in space.

Could the fictional Captain James T. Kirk, commander of the *Starship Enterprise* and later a Star Fleet admiral, be the model for our future leaders in space? Kirk, like Washington, is a leader and a military man. But he is also a person with a broad background and varied interests. In most "Star Trek" adventures, Kirk meets the challenges he faces by using his mental skills rather than force.

According to the "Cadet's Orientation Handbook" of the "Star Trek Role Playing Game," some of Kirk's greatest skills are in starship combat and leadership, but he is also an expert on the life of Abe Lincoln, negotiation and diplomacy, and federation law.

While there has yet to be an episode of "Star Trek" in which Kirk leaves Star Fleet, we can easily imagine him governing a planet or even the federation itself. Could a Kirk-like figure become the "father of our society" in a star system far from Earth?

which have less gravity than Earth, the body will adapt in ways that we cannot yet predict.

It is also possible, even though it would take a very long time, that a new species would evolve in space. This species would be so different from humans that we could no longer mate and have offspring.

These changes will increase the differences between people living on Earth and those living in outer space. Today's diplomats still struggle to create unity among people on Earth. The people of tomorrow will need to find common ground among humans spreading out into the solar system.

These are only some of the challenges that will be faced by the "people of tomorrow."

WHO ARE THE "ALIENS"?

Imagine this: an inhabitant of the third planet from a star not far from Earth crouches over an instrument that looks very much like a television set.

She looks like a human being, except that she is taller than most humans. Her planet is smaller than

Figure 34—It is possible that, in the distant future, human beings who live in outer space will adapt to their weightless environment and look very different from their ancestors on earth.

Earth, and exerts a lesser gravitational pull on living things, so they are able to grow more. Her skin is quite dark, because the sun in her system is just a bit brighter than ours. Her name is A-Tae and she is a scientist, in charge of her institute's SILU (Search for Intelligent Life in the Universe) project. She has been looking twenty years, without success. Now she is getting something. Her excitement is enormous.

She twists the dials, trying to bring in a picture that fades and breaks up, then fades again. She struggles to hold it, and it stabilizes. She stands away from the instrument and looks at the picture in awe.

On the screen, there is a group of young humanoids all dressed alike. They seem normal, by galactic standards, except that they have enormous *black* ears on top of their heads. They are singing a song that she can just barely make out:

Who's the leader of the club that's made for you and me?
M-I-C-K-E-Y-M-O-U-S-E!

A-Tae rushes from the room, seeking her colleagues. She has made the most momentous discovery of her species. She has found intelligent life on another planet, and has apparently witnessed one of their religious ceremonies!

Does that sound a bit farfetched to you? Well, it was meant to be humorous, just to make a point. If there are other intelligent beings living on other planets in the galaxy, they are probably looking for us, even as we are looking for them. Our television and radio signals have been radiating out into space for over fifty years now, and someone could be picking them up.

We are listening for the E.T.'s, and we are also

sending manned and unmanned missions out into space. If they are out there, we should expect them to be doing the same. Eventually, contact is likely to take place between them and us.

When that happens, it will be another of those moments that changes the human view of space forever. At the moment, we are faced with many unknowns. We do not know for certain whether there are many planets outside our solar system. We also do not know whether planets are even necessary for intelligent life to evolve. If there are planets circling other stars, and if life is evolving there, we do not know if the evolutionary process will be the same as it is here on Earth.

We do know a few things, though. We know that there are about a billion stars like our sun, in our galaxy alone. Scientists have a lot of evidence that many, or most stars, do have planets around them. So far, it appears that matter, energy, and the building blocks of life behave pretty much the same throughout the universe.

We are a bit like the jury in a difficult court case. We cannot yet prove anything, but all the circumstantial evidence points to a universe full of life and intelligence. Of one thing we can be sure: when contact comes, it will be as dramatic as any of those other moments when our views of space changed forever.

If we make contact with other species, we will be "aliens" to them and they will be "aliens" to us. If their culture is older than ours, then our entire future will change. Instead of finding our own way into the universe, we will have instructors and guides. If our culture is older than theirs, we may become the teachers.

We come from a planet where human beings with different skin coloring have difficulty getting along. Can you imagine the challenge we face in dealing with beings who are not human? Perhaps it will help us to see how superficial our own differences really are!

REVIEW QUESTIONS

1. Why do you think that the "Star Trek vision" appeals to people?
2. How is today's situation on the space frontier similar to the situation on the American frontier? How is it different?
3. Do you think large numbers of people will be able to live in outer space? Why or why not?
4. Why would people living in a weightless environment begin to look different from people on Earth?
5. Why do we believe that there are other intelligent beings in our galaxy and beyond? What do you think?
6. Would you like to live off the planet Earth? Why or why not?

6 Epilogue

A GLANCE AT THE YEAR 2010

Anne Morito-Jones floated over to the viewport and looked out. Behind her, she could see the beautiful Earth floating too, in an inky blackness punctuated by the light of millions of stars. In her mind, Anne said goodbye to Mother Earth with a mixture of sadness and joy. She was sixteen years old, and her family had decided, after many long discussions, to migrate to the new international settlement on the moon, Luna II.

Yes, Anne missed the Earth, but she looked forward to learning more about the moon. She had heard that the kids there were really different. Anyway, Anne already felt sorry for her friends back in Columbus, Ohio, who still had to put up with gravity!

As Anne gazed back at the Earth growing increasingly smaller in the viewport, she found herself thinking back to something she had read not so long ago about exploration. She saw the image of a man dressed in furs staring across an icy stretch of water as a caribou crossed to the other side. As the image faded, she knew, just for a moment, how that man, her distant ancestor, must have felt as he picked up his spear and followed the caribou into the unknown.

Appendices

Looking at Careers in Space

Careers in the Space Field Today

Of all the human beings who have ever lived on the planet Earth, you have a unique choice today: you can decide to become involved in the human exploration of outer space, or to ignore it. You may even get to choose whether you want to live on the Earth or somewhere else in the solar system.

There is no right or wrong choice. It isn't better to live on a space settlement or to live on Earth. However, it will be hard to avoid thinking about it. As human beings put more pressure on Earth's ecological systems, everyone will be asking, "What shall we do?" Many of the decisions will involve the use of space technology. There will also be a debate over whether expansion into space is a good solution, or whether we should simply limit our growth here on Earth.

Your choice is really about whether to be involved or just be affected by other people's decisions. Let's say you want to get involved. What can you do? Well, we have already pointed out several ways to become active, such as joining a space-oriented organization. But let's suppose you *really* want to get involved. Are there careers in the space field today and tomorrow?

In fact, there are many careers available to you in the space field today. You can, of course, work for a national space program, such as the National Aeronautics and Space Administration (NASA) in many different areas. When young people think of NASA, they naturally think about becoming an astronaut. Being an astronaut certainly is a career, and NASA selects new candidates every couple of years. The requirements are demanding, including a college degree in a field specified by NASA as desirable.

At the moment, the American astronaut corps is relatively small, about one hundred people. However, it is likely to grow, especially as the United States begins work on the permanently manned space station. As space exploration and development becomes a bigger field, so will the need for astronauts. It's a little bit like the need for pilots in 1928, as compared to 1988. In the early days, only a few highly skilled pilots were needed, but now thousands are required in the commercial aviation industry alone.

The astronaut corps gets most of the publicity, but there are many other jobs in NASA. You might, for example, work as an engineer designing new spacecraft or propulsion systems. If you are interested in how people work together, you could become involved in "human factors" studies. If you are intrigued by communications, you could be a public affairs officer, helping NASA explain its programs to the public. If extraterrestrial life excites you, then consider getting involved in NASA's SETI program.

Beyond NASA, there are the many companies that work under contract to the space agency. These companies need many of the same skills that NASA does, and others besides. Then, there are new companies emerging every day that are trying to find better ways to commercialize the space environment. If you are innovative and creative, you may want to get involved with one of these new "start-ups."

There are other ways to work in the space field indirectly. For example, you might want to become a teacher, using different aspects of space exploration in your classes. You might also become a writer specializing in space-related subjects, or a lawyer focusing on space law. There are very few professions that cannot be connected to space exploration and development in some way.

Careers in the Space Field Tomorrow

In the future, as people begin to live in space, we will need all the same professions there that we need on Earth.

People living on the moon or in space settlements will be cooks, farmers, doctors, lawyers, plumbers, mechanics, child care experts, etc.

Today, life in space is in the earliest pioneering stages. Only a few hardy souls venture out onto that demanding frontier. However, many people have a vision of life in space changing. They see people working and raising their families on that frontier. They see new wealth being created and new societies being formed. If that vision is to be realized, it will take all kinds of people, not just astronauts and cosmonauts.

There will also be new kinds of careers on the space frontier, such as "lunar metals expert" and "extraterrestrial psychologist."

Of course, it will take time—years and decades perhaps. However, at least part of this vision may be realized within your lifetime, or that of your children.

In the space field of tomorrow, just think of the kind of career you've always wanted on Earth and imagine it happening "out there." That's the starting point in planning a career in the space field of tomorrow.

If you have wanted to be involved in space exploration and development, but have felt that you could not make a contribution, think again. Space exploration is not just a technological achievement. It is also a great human achievement, and it will require all kinds of people to help make it happen.

In the words of the president's National Commission on Space: "Settling North America required the sustained efforts of laborers and farmers, merchants and ministers, artisans and adventurers, scientists and seafarers. . . ."

In the same way, settling the space frontier will require the sustained efforts of a lot of people just like you.

Organizations Involved in the Exploration and Development of Outer Space

Challenger Center: The Challenger Center for Space Science Education was founded by the families of the astronauts who died in the space shuttle *Challenger* disaster. The center is committed to "completing the mission" of the *Challenger*, which was carrying the first teacher in space. The center focuses on creating a realistic learning environment for young people to learn more about space exploration.

For information: Challenger Center, P.O. Box 96306, Washington, D.C. 20077-7198

National Aeronautics and Space Administration (NASA): Thirty years after its founding, NASA remains a focal point for much of America's space exploration efforts. The agency is also a good source of information about current space-related activities and careers in the space field.

For general information: National Aeronautics and Space Administration, Public Affairs Office, Washington, D.C. 20546. For information on astronaut qualifications: National Aeronautics and Space Administration, Lyndon B. Johnson Space Center, Houston, Tex. 77058

National Space Society (NSS): NSS is part of a "family of organizations" with headquarters in Washington, D.C., working on space-related issues. It is also a grassroots activist organization with local chapters nationwide. The society holds an annual space development conference, publishes a magazine, and alerts members to important developments in the space field. Chapters hold meetings at a local level and conduct a variety of projects.

For information: National Space Society, 922 Pennsylvania Avenue, S.E., Washington, D.C. 20003

Planetary Society: Headed by astronomer Carl Sagan, the society supports unmanned and manned exploration of space, and has been a leader in urging a joint U.S.-Soviet mission to Mars. The society also publishes a magazine and holds conferences.

For information: Planetary Society, 65 North Catalina Avenue, Pasadena, Calif. 91106

Space Studies Institute (SSI): SSI's research is aimed at supporting large-scale human settlement of space, utilizing extraterrestrial materials from the moon and asteroids. The emphasis is on "bootstrapping" and private initiatives. Founded by Dr. Gerard K. O'Neill, SSI holds a biennial conference and publishes a monthly newsletter.

For information: Space Studies Institute, P.O. Box 82, Princeton, N.J. 08540

United States Space Camp and Space Academy: Space Camp offers five-day and ten-day educational experiences for young people, culminating in simulated space shuttle "missions" on the last day of camp. There is also a three-day "adult space camp." Enrollment has grown from around 700 in 1982 to some 18,000 in 1988. The original Space Camp is in Huntsville, Alabama, but the concept is being duplicated in other areas of the country.

For information: United States Space Camp, 1 Tranquility Base, Huntsville, Ala. 35807-0287

Students for the Exploration and Development of Space (SEDS): SEDS is a student-run organization. Its purpose is to support and channel student interest in space exploration and development. There are a number of SEDS chapters on college campuses around the country.

For information: SEDS, Massachusetts Institute of Technology, W20-445, Cambridge, Mass. 02139

International Space University (ISU): ISU offers a curriculum focusing exclusively on space-related issues to an international body of students. It is currently running summer sessions at different universities around the world. The ultimate goal: a campus in orbit.

For information: International Space University, 636 Beacon Street, Suites 201-202, Boston, Mass. 02215

Institute for Security and Cooperation in Outer Space (ISCOS): ISCOS promotes international cooperative space activities as an alternative to deployment of space weapons. ISCOS activities include research, activism, and outreach.

For information: ISCOS, 8 Logan Circle, Washington, D.C. 20004-3727

Space Frontier Foundation: A grassroots organization supporting the opening up of the space frontier, starting with a return to the moon.

For information: The Space Frontier Foundation, c/o The Intrepid Sea-Air-Space Museum, 46 Street at 12th Avenue, New York, N.Y. 10036

The Space Age Begins

Partial Text of the Announcement by the Soviet News Agency, TASS, of the Launching of *Sputnik I*, October 4, 1957

For several years research and experimental designing work has been under way in the Soviet Union to create artificial satellites of the earth. It has already been reported in the press that the launching of the earth satellites in the U.S.S.R. was planned in accordance with the program of the International Geophysical Year.

As a result of the intensive work by research institutes and designing bureaus the first artificial earth satellite in the world has now been launched. The first satellite was successfully launched in the U.S.S.R. on Oct. 4.

First Man in Space

Excerpts From Statement by Soviet Union's TASS News Agency on Flight of Yuri Gagarin, April 12, 1961, *New York Times*

The world's first satellite spaceship *Vostok*, with a man on board, was placed in an around-the-world orbit in the Soviet Union on April 12, 1961.

The pilot space navigator of the *Vostok* is a citizen of the U.S.S.R., Flight Maj. Yuri Alekseyevic Gagarin.

The launching of the multi-stage space rocket was successful, and after attaining the first escape velocity and the separation of the last stage of the carrier-rocket, the spaceship went into free flight on an around-the-world orbit. . . .

Space Navigator Gagarin withstood satisfactorily the placing of the satellite ship *Vostok* into orbit, and at present feels well. The systems insuring the necessary vital conditions in the cabin of the spaceship are functioning normally.

The spaceship, with Navigator Gagarin on board, is continuing its flight in orbit.

Commitment to the Moon

Press Reports on President Kennedy's Speech to Congress Asking for Funds to Accelerate the Space Program, *New York Times*, May 25, 1961

Washington, May 25 (by W. H. Lawrence)—President Kennedy today proposed to Congress bold and expensive new measures to rocket a man to the moon, to expand non-nuclear military strength and to increase foreign aid spending.

These actions, he said, are needed to promote a "freedom doctrine" around the globe. . . .

To emphasize the urgency of his proposals, Mr. Kennedy appeared personally before a joint session of Congress. His speech was televised and broadcast nationally by all networks. . . .

The president's major emphasis was on the necessity of getting a man to the moon, and of getting him there first, if possible. He told Congress a firm national decision was essential now on whether this nation would go all the way on a big space program that "will last for many years and carry very heavy costs."

"If we were to go only half way to reduce our sights in the face of difficulty, it would be better not to go at all," he said.

The president conceded that Congress might not hold the same views he did. But "I believe that we should go to the moon," he declared. It was time, he said, for the United States "to take a clearly leading role in space achievement," which may "hold the key to our future on earth."

America Rejoices

Glenn Goes into Orbit (Excerpts From *New York Times* Report of February 21, 1962)

Cape Canaveral, Fla., Feb. 20 (by Richard Witkin)—John H. Glenn, Jr. orbited three times around the earth today and landed safely to become the first American to make such a flight.

The 40-year-old Marine Corps lieutenant colonel traveled about 81,000 miles in 4 hours 56 minutes before splashing into the Atlantic at 2:43 P.M. Eastern Standard Time.

He had been launched from here at 9:47 A.M.

The astronaut's safe return was no less a relief than a thrill to the Project Mercury team, because there had been real concern that the *Friendship 7* capsule might disintegrate as it rammed back into the atmosphere.

There had also been a serious question whether Colonel Glenn would complete three orbits as planned. But despite persistent control problems, he managed to complete the entire flight plan. . . .

The Goal Is Achieved

Apollo 11 Lands on the Moon. (Excerpts from Transcript of Dialogue Between "Eagle" Lunar Module and Houston Mission Control, *New York Times*, July 21, 1969)

(Note: Some of the information in parentheses was supplied by the *New York Times*. Some of it was supplied by the authors of this book, for clarity.)

EAGLE (the lunar module): Houston, Tranquility Base here. The Eagle has landed.

HOUSTON: Roger, Tranquility, we copy you on the ground. You've got a bunch of guys about to turn blue. We're breathing again. Thanks a lot.

TRANQUILITY BASE: Thank you.

HOUSTON: You're looking good here.

TRANQUILITY BASE: A very smooth touchdown.

HOUSTON: Eagle, you are stay for T1 [the first step in the lunar operation]. Over.

TRANQUILITY BASE: Roger. Stay for T1.

HOUSTON: Roger and we see you venting the ox [oxygen].

TRANQUILITY BASE: Roger.

COLUMBIA (the command and service module): How do you read me?

HOUSTON: Columbia, he has landed Tranquility Base. Eagle is at Tranquility. I read you five by [I read you well]. Over.

COLUMBIA: Yes, I heard the whole thing.

HOUSTON: Well, it's a good show.

COLUMBIA: Fantastic.

TRANQUILITY BASE: I'll second that.

APOLLO CONTROL: The next major stay-no stay will be for the T2 event. That is at twenty-one minutes, twenty-six seconds after initiation of power descent.

COLUMBIA: Up telemetry reset to reacquire on high gain.
HOUSTON: Copy. Out.
APOLLO CONTROL: We have an unofficial time for that
touchdown of one hundred and two hours, forty-five
minutes, forty-two seconds, and we will update that.

The Earth Reacts

Opinions of People Interviewed After the Moon Landing (*New York Times*, July 21, 1969)

Jim Aiello (a 12-year-old Michigan boy): I read in books that ever since early man they've wondered what the moon was made out of and why it's up there. Now we're going to find out why. It's just like Columbus—finding a new world. People landing on the moon are just like another Columbus. . . .

Charles Lindbergh (first man to fly the Atlantic alone): Along with this development of space, which is really a flowering of civilization toward the stars . . . we must protect the surface of the earth. That is ever more important. Our environment on the surface is where man lives. . . .

George Wald (biologist): . . . let us be glad to explore this corner of space, the solar system, the more to feel at home in it; the more to realize that our home in space is the earth; and the more to treat it as our home, rather than a fief to exploit and consume. . . .

Lewis Mumford (historian and expert on cities): [The moon landing] is a symbolic act of war, and the slogan the astronauts will carry, proclaiming that it is for the benefit of mankind, is on the same level as the Air Force's monstrous hypocrisy—"Our Profession Is Peace. . . ."

Paul Goodman (poet and novelist): From the first *Sputnik*, I have been enchanted by this adventure of Man . . .

The beautiful equations of Galileo, Kepler, and Newton—and naturally one thinks also of Columbus and Magellan . . . It's good to "waste" money on such a moral and esthetic venture. There are our cathedrals. . . .

Ella Winter (author): That man can achieve this means we could do almost anything on earth if only we'd do it. . . .

David Reisman (sociologist): The possibility of nuclear destruction has made a greater difference in my life than space exploration is likely to make . . . And when I think about the moon landing in international terms, I see it in one aspect as a moral equivalent of war . . . a bit like the Olympics. . . .

René Dubos (microbiologist): There are men in the Apollo capsules, but what they do has all been simulated and is controlled at every step by an incredibly efficient technology—the dreaded portent of what could happen in our lives and our thoughts in the world to come. . .

Pablo Picasso (artist): It means nothing to me. I have no opinion about it and I don't care.

Jean Monnet (economist) I am for it. It makes me grow beyond myself. It breaks down the last frontier. . . .

R. Buckminster Fuller (architect and inventor) The landing on the moon is in the dead-center of evolutionary events. . . .

A New Vision

Mission Statement and Rationale for Exploring and Settling the Solar System. From "Pioneering the Space Frontier," Report of the National Commission on Space, 1986

A PIONEERING MISSION FOR TWENTY-FIRST CENTURY AMERICA

To lead the exploration and development of the space frontier, advancing science, technology, and enterprise, and building institutions and systems that make accessible vast new resources and support human settlements beyond Earth orbit, from the highlands of the moon to the plains of Mars.

RATIONALE FOR EXPLORING AND SETTLING THE SOLAR SYSTEM

Our Vision: The Solar System as the Home of Humanity
The solar system is our extended home. Five centuries after Columbus opened access to "The New World" we can initiate the settlement of worlds beyond our planet of birth. The promise of virgin lands and the opportunity to live in freedom brought our ancestors to the shores of North America. Now space technology has freed humankind to move outward from Earth as a species destined to expand to other worlds.

Our Purpose: Free Societies on New Worlds
The settlement of North America and other continents was a prelude to humanity's greater challenge: the space frontier. As we develop new lands of opportunity for ourselves

123

and our descendants, we must carry with us the guarantees expressed in our Bill of Rights: to think, communicate, and live in freedom. We must stimulate individual initiative and free enterprise in space.

Our Ambition: Opening New Resources to Benefit Humanity
Historically, wealth has been created when the power of the human intellect combined abundant energy with rich material resources. Now America can create new wealth on the space frontier to benefit the entire human community by combining the energy of the sun with materials left in space during the formation of the solar system.

Our Method: Efficiency and Systematic Progression
In undertaking this great venture we must plan logically and build wisely. Each new step must be justified on its own merits and make possible additional steps. American investments on the space frontier should be sustained at a small but steady fraction of our national budget.

Our Hope: Increased World Cooperation
In his essay, "Common Sense," published in January of 1776, Tom Paine said of American independence, " 'Tis not the affair of a City, County, a Province, or a Kingdom; but of a Continent. . . . 'Tis not the concern of a day, a year, or an age; posterity are virtually involved in the contest, and will be more or less affected to the end of time, by the proceedings now." Exploring the universe is neither one nation's issue, nor relevant to our time. Accordingly, America must work with other nations in a manner consistent with our Constitution, national security, and international agreements.

Our Aspiration: American Leadership on the Space Frontier
With America's pioneer heritage, technological preeminence, and economic strength, it is fitting that we should lead the people of this planet into space. Our leadership role should challenge the visions, talents, and energies of

young and old alike, and inspire other nations to contribute their best talents to expand humanity's future.

Our Need: Balance and Common Sense
Settling North America required the sustained efforts of laborers and farmers, merchants and ministers, artisans and adventurers, scientists and seafarers. In the same way, our space program must combine with vigor and continuity the elements of scientific research, technological advance, the discovery and development of new resources in space, and the provision of essential institutions and systems to extend America's reach in science, industry, and the settlement of space.

Our Approach: The Critical Lead Role of Government
As formerly on the western frontier, now similarly on the space frontier, government should support exploration and science, advance critical technologies, and provide the transportation systems and administration required to open broad access to new lands. The investment will again generate in value many times its cost to the benefit of all.

Our Resolve: To Go Forth "In Peace for All Mankind"
When the first Apollo astronauts stepped onto the moon, they emplaced a plaque upon which were inscribed the words, "We came in peace for all mankind." As we move outward into the solar system, we must remain true to our values as Americans: To go forward peacefully and to respect the integrity of planetary bodies and alien life forms, with equality and opportunity for all.

Glossary

Apollo Program: The American space program that fulfilled the goal of landing a man on the moon and bringing him back safely by the end of the decade 1960–69.

Apollo/Soyuz: A joint American/Soviet space flight conducted in Low Earth Orbit in 1975.

Artificial Gravity: The creation of a gravitational force by rotating a spaceship or space station.

Ascent Engine: The engine used to power the lunar module of an Apollo spacecraft from the surface of the moon to a rendezvous in orbit with the Command and Service Module.

Astronaut: A person who has passed the requirements to fly in space, as defined by a recognized national space program; one who has flown in space, defined as a region fifty miles above the surface of the Earth; in particular, an American space voyager.

Cape Canaveral: The location in Florida from which most manned and unmanned American space missions have been launched. "The Cape" was briefly renamed Cape Kennedy, but the name was later changed back to Canaveral, with the NASA installation retaining the name "Kennedy Space Center."

Cold War: The name for the conflict between the United States and the Soviet Union, in which no shots were fired, but the two countries competed for dominance, especially in the "Third World."

Command and Service Module: That part of an Apollo spacecraft in which the crew lived on the way to and from the moon, and which remained in lunar orbit while the lunar module went to the surface.

Communications Satellite: A satellite used for transmitting information from one part of the Earth to another.

Copernican Revolution: The change in thinking that occurred when people began adopting the view of Copernicus that the Earth and the other planets revolved around the sun.

Cosmology: The study of the nature and structure of the universe.

Cosmonaut: The same definition as that of an astronaut, except that it refers specifically to those trained by the Soviet Union.

Declaration for Space: Initial section of *Pioneering the Space Frontier*, the report of the National Commission on Space.

Epicycle: A looping pattern that early astronomers devised to describe the apparent movement of planets in their orbits around the Earth.

European Space Agency (ESA): A group of twelve European nations, plus Canada, which jointly fund and coordinate space projects, such as the development of the *Ariane* rocket.

Explorer: The first successful satellite launched by the United States. It went into orbit in early 1958.

Gemini Program: The American space program that followed the Mercury program and laid the groundwork for the Apollo missions to the moon.

Global Village: A term coined by media expert Marshal McLuhan to describe the effect of planetwide communications capabilities on the relationships between people on Earth.

International Geophysical Year (IGY): A program of international cooperation that provided the framework in which the first artificial satellites, including *Sputnik I*, were launched.

Jupiter: The rocket used by Wernher Von Braun and his team as the basic component of the vehicle used to launch the *Explorer* satellite. The entire system was known as the *Juno* rocket.

Lagrange Points: Named for their discoverer, LaGrange

points are areas between planetary bodies where the gravitational pull of the two bodies is most balanced. This makes LaGrange points good places to locate structures such as space habitats.

Law of Gravity: A law of nature enunciated by Newton that describes how physical bodies, such as two planets, or a planet and a star, will be attracted to one another.

Laws of Motion: Laws of nature, also discovered by Newton, that describe how bodies at motion in space will behave under different circumstances.

Low Earth Orbit: A region approximately one hundred to two hundred miles above the Earth's surface where many space missions take place and where all of the conditions of the space flight experience, such as weightlessness, are present.

Lunar Module: That part of the Apollo spacecraft that detaches from the Command and Service Module and descends to the surface of the moon.

Mariner: A series of unmanned spacecraft that were sent to Mars by the United States in the 1960s and 1970s.

Milky Way Galaxy: A collection of stars functioning together as a system, of which our sun is a member. Called the Milky Way because of its appearance when we look toward the galactic center from our location toward one edge of the galaxy.

Mir (Peace): The permanently manned Soviet space station now in Low Earth Orbit.

National Aeronautics and Space Administration (NASA): The agency founded in 1958 to coordinate all American civilian space activities.

National Commission on Space (NCOS): A group of prominent Americans appointed in 1985 by President Reagan to develop a fifty-year plan for the American civilian space program.

New World: A name given to the North and South American continents by Europeans during the previous great age of exploration in the fifteenth, sixteenth, and seventeenth centuries.

O-Ring: The component that failed during the flight of the space shuttle *Challenger* in 1986, resulting in the destruction of the spacecraft.

Pioneer: Two unmanned spacecraft launched by the United States in the early 1970s to explore the outer solar system. *Pioneer 10* and *11* were the first manmade objects to leave the solar system.

Remote Sensing Satellites: Satellites that look back at the Earth in order to do mapping, take photographs, locate natural resources, etc.

Saturn V: The rocket that took astronauts to the moon during the Apollo program.

Sea of Tranquility: Landing site for the *Apollo 11* mission. Not actually filled with water, it was so named by early astronomers because it looks like a sea from the Earth.

SETI: The Search for Extraterrestrial Intelligence, usually referring to the use of receiving equipment to detect radio signals from advanced technical civilizations in other parts of the galaxy or in other galaxies.

Soyuz: Soviet spacecraft first used in the early 1970s, and still used today to take cosmonauts to the Soviet space station in Low Earth Orbit.

Space Shuttle: A reusable spacecraft designed for trips to Low Earth Orbit. The first successful space shuttle was launched by the United States in 1981.

Space Walk: The leaving of a spacecraft by an astronaut or cosmonaut to maneuver in free space. Also known as an EVA, or extravehicular activity.

Skylab: The first American space station, occupied by three crews in the early 1970s.

Sputnik I: The first artificial Earth-orbiting satellite, launched by the Soviet Union in 1957.

Sputnik II: The second artificial satellite launched by the Soviet Union; it carried with it a dog named Laika.

Strategic Defense Initiative (SDI): A proposal made by President Reagan in 1983 to construct a defensive shield against nuclear missiles in Earth orbit.

Suborbital Flight: A flight that does not take a rocket or its payload into orbit. The first American manned space-flights were suborbital.

V-2 Rocket: Rockets developed by Wernher Von Braun and his colleagues for use by the Germans during World War II; became the basis for most American rocket designs after the war.

Vanguard: The original American satellite program. Initially a failure, Vanguard did succeed in launching a satellite in 1958.

Viking: An unmanned probe that the United States sent to Mars in 1976. One Viking spacecraft landed while the other orbited the planet.

Voyager: Two American spacecraft launched in 1977 to make close encounters with the four planets of the outer solar system.

Zero-G: Zero gravity, or the state of weightlessness caused by moving outside the gravitational pull of a planetary body such as the Earth.

Bibliography

Allen, Joseph P. *Entering Space: An Astronaut's Odyssey*. New York: Stewart, Tabori & Chang, 1984.

Carpenter, M. Scott *et al. We Seven: By the Astronauts Themselves*. New York: Simon and Schuster, 1962.

Collins, Michael L. *Carrying the Fire*. New York: Farrar, Straus and Giroux, 1974.

DeWaard, E. John and Nancy. *History of NASA: America's Voyage to the Stars*. New York: Exeter Books, 1984.

Furniss, Tim. *Space Flight: The Records*. Guiness Superlatives, Ltd., Enfield, Middlesex, England, 1985.

Gagarin, Yuri and Lebedev, Vladimir. *Survival in Space*. New York: Frederick Praeger, 1969.

Hoffman, Jeffrey A. "An Astronaut's Diary," Audiotape, Washington, D.C.: National Public Radio, 1985.

Holmquist, Michael. "The Philosophical Bases of Soviet Space Exploration." In *The Key Reporter* (Winter 1985–86.)

Lewis, Richard S. *Appointment on the Moon*. Ballantine Books, New York, 1968.

Mason, Robert Grant (ed.). *Life in Space*. Arlington, Va.: Time-Life Books, 1983.

McDougall, Walter A. . . . *the Heavens and the Earth: A Political History of the Space Age*. New York: Basic Books, Inc., 1985.

National Commission on Space. *Pioneering the Space Frontier*. New York: Bantam Books, 1986.

Pardoe, Geoffrey K. C. *The Future for Space Technology*. London and Dover, N.H.: Frances Pinter (Publishers), 1984.

United States Department of Commerce. *Space Commerce: An Industry Assessment*. Washington, D.C.: May 1988.

White, Frank. *The Overview Effect: Space Exploration and Human Evolution*. Boston: Houghton Mifflin, 1987.

Wolfe, Tom. *The Right Stuff*. New York: Farrar, Straus, Giroux, 1979.

NOTE

Some books and articles were especially useful to the authors in writing this book, and they deserve special mention beyond simply being included in the bibliography.

These include:

Survival in Space by Gagarin and Lebedev, for Yuri Gagarin's quote regarding his feelings while in space.

Life in Space, edited by Robert Grant Mason, for many details of the early space program and its descriptions of the Mercury and Apollo programs in particular.

NASA: America's Voyage to the Stars, by E. John and Nancy DeWaard, for details on both the manned and unmanned space programs.

. . . the Heavens and the Earth: A Political History of the Space Age, by Walter A. McDougall, for details of the political environment just before and just after *Sputnik.*

"The Philosophical Bases of Soviet Space Exploration" by Michael Holmquist, for information about Tsiolkovsky and Fyodorov.

Index